Tune, Rebuild or Modify
ROCHESTER CARBURETORS
by Doug Roe

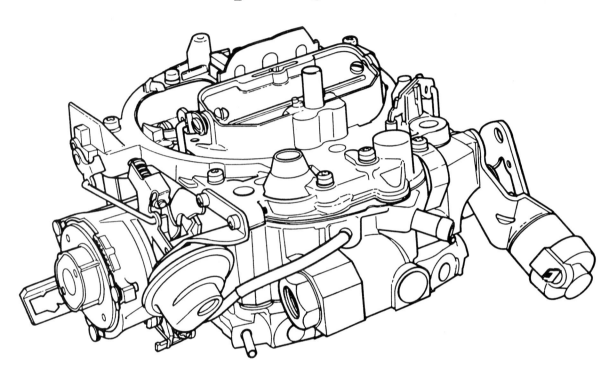

Photos: Doug Roe, Bill Fisher, others noted
Cover photo: Bill Keller

HPBooks
are published by
The Berkley Publishing Group
200 Madison Avenue, New York, NY 10016
ISBN: 0-89586-301-4   Library of Congress Catalog Card Number 86-81204
© 1981 Price Stern Sloan, Inc.
Printed in the U.S.A.
Revised edition

25   24   23   22   21   20   19

High-performance carburetion, page 110.

Multiple-carburetor manifolds, page 106.

——Acknowledgments——

Special thanks go to many who probably are unaware of their contribution to this book. They are fellow workers, supervisors and others who provided me with challenging problems and great direction; consequently, an understanding of carburetors and related vehicle components. Regretfully, the majority can't be listed, but if in the past you can recall such occasions—thanks. To those who helped and won't read these words, there will always be recognition and a quiet thanks.

Of recent acquaintance, there are Al Sears of Rochester Products Division test management, Sturdy Thomas of Digital Automotive Systems and Glen Grissom of HPBooks who deserve thanks for their direct input. Tom VanGriethuysen and Jonathan Gugel of Rochester Products Division, GM supplied photos and drawings. John Adrian helped with disassembing and reassembling ECM-controlled carburetors for photographs. Jim Le Roy helped with the formulas and related terms.

It's always appreciated when family and friends tolerate the constant paper and picture mess scattered throughout home and office as a book is being written. And thanks to all the automotive editorial staff at HPBooks for assisting and exhibiting patience when either or both were needed.

Emissions control evolution, page 169.

Servicing Q-jet choke, page 81.

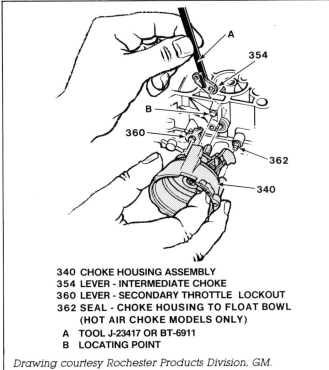

340 CHOKE HOUSING ASSEMBLY
354 LEVER - INTERMEDIATE CHOKE
360 LEVER - SECONDARY THROTTLE LOCKOUT
362 SEAL - CHOKE HOUSING TO FLOAT BOWL
    (HOT AIR CHOKE MODELS ONLY)
  A TOOL J-23417 OR BT-6911
  B LOCATING POINT

Drawing courtesy Rochester Products Division, GM.

CONTENTS

Introduction

Rochester Products Division (RPD) of General Motors has long been a leader in designing and manufacturing carburetors. RPD prints manuals, training books, and parts reference guides by the thousands, and independent technical writers contribute to this vast library, too. Many of these publications cover general service and maintenance on Rochester carburetors.

I wrote this book to offer you knowledge beyond that in the typical service manual; much of it has never been presented elsewhere. This extensive revision keeps the basic chapters about theory of carburetion and engine air/fuel requirements of previous printings. It now includes new information on electronic controls for Rochesters, Quadrajet (Q-jet) service, emission systems, high-performance modifications, and tuning for performance or economy. It also describes the versatility of these excellent carburetors for street and racing applications.

Photos and illustrations throughout detail design features and operation. Many show how special parts are used to improve performance in specific applications and how stock parts should be checked or blueprinted to serve you best.

This book isn't a substitute for factory service and overhaul manuals that supply basic tuning information for certain vehicles or extensive repair information for specific carburetors. It doesn't have complete adjustment details for Rochester carburetors, because these vary from vehicle to vehicle. And, with changing emission requirements, each automobile's carburetor is factory-adjusted for optimum emission control.

The carburetor is often the least-understood and most-worked-on part on a car. I have more than three decades of experience with them and I'm still learning.

My involvement started in the early '50s with Stromberg Carburetor, a division of Bendix Aviation, with four years at their Elmira, N.Y. Eclipse Plant, working on road and laboratory testing. I then worked

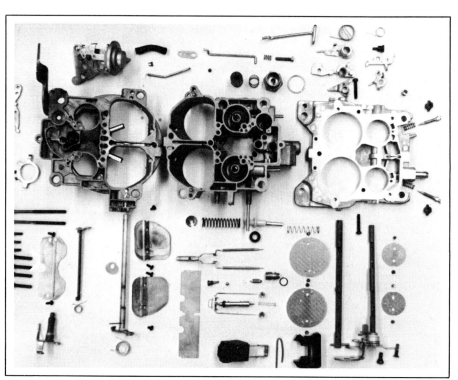

Got your Rochester Quadrajet (Q-jet) apart and can't quite get it together? No problem, read on and learn how it works and what modifications improve its performance.

four years with RPD of General Motors in their test laboratory and as a field representative to Chevolet Motor Division in Michigan. A transfer to Chevrolet Motor Division supplied 12 more years of concerted effort in engine/carburetor development and test programs.

To further supplement my research, I have interviewed many automotive people, including GM training instructors, to help you better understand carburetion in general, and Rochester carburetors, in particular.

TECHNICAL IDEAS INTO WORDS

One difficulty in writing a technical book is presenting theory and practical information in understandable language. Consequently, some of the words and ideas will be familiar to some and foreign to others. For example, the word *signal* means to a carburetion engineer the amount of pressure drop in the venturi, which signals the fuel supply to meter the correct amount of gasoline.

If you initially meet some unfamiliar terms and concepts, please be patient. They'll be explained and sorted with definitions, some calculations, descriptions, practical examples, drawings and detailed photos.

How Your Carburetor Works 1

Understanding how a carburetor works is an important step in extracting top performance from an engine/carburetor combination. Once aware of what happens inside a carburetor and how it works with the engine, you should be able to service, modify, or change it with confidence. You won't have to rely on self-proclaimed experts or rumors about the current "hot setup."

There's nothing tricky about a carburetor. No black magic makes one work differently from another—all types operate essentially the same, regardless of the number of barrels or venturis. Of course there are differences, but these are related to how control systems are built in a particular carburetor.

Carburetors are straightforward devices. Just as four-cycle engines—from one-cylinder lawn mowers to high-powered V8s—operate on the same principle of intake, compression, power and exhaust, carburetors operate on universal principles.

Even as you could learn about an engine by examining the workings of one cylinder, let's begin by scrutinizing a one-barrel carburetor. Following this chapter in sequence will help you grasp the fundamentals of a one-barrel and understand the discussions about more complex two- and four-barrel carburetors.

The carburetor does several important jobs:

● Controls airflow to the engine to control power output.

● Mixes air and fuel in correct proportions for engine operation.

● Aids vaporization of air/fuel mixture and, in part, puts it in a homogeneous state for combustion.

To satisfy the range of requirements of engines and drivers, several basic carburetor systems are used. These are:

● Inlet.
● Main.
● Power.
● Accelerator pump.
● Idle.
● Choke.

We'll examine each of these, as well as secondary throttle operation and other carburetor functions.

INLET (FLOAT) SYSTEM

The inlet system has three primary components:

● Fuel bowl.
● Float.
● Fuel inlet valve (needle and seat).

Fuel for the carburetor's metering is stored in the *fuel bowl*. The *float* maintains

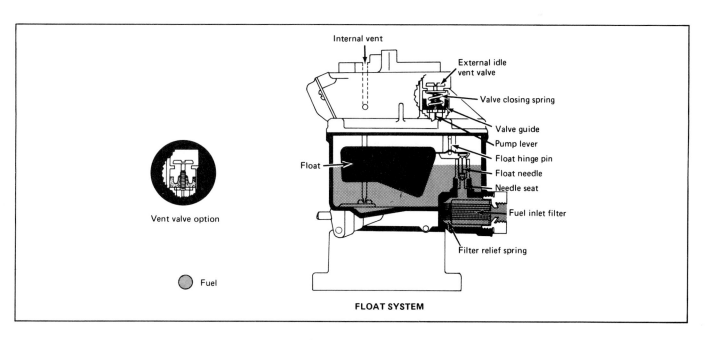

Vent valve option

Fuel

FLOAT SYSTEM

the specified fuel level in the fuel bowl because the correct mixture is delivered only when fuel is at this level. The fuel level affects *fuel handling,* which is the carburetor's metering ability during vehicle maneuvers—acceleration, turns and stops.

OPERATION

The amount of fuel entering the bowl through the *fuel inlet valve (needle valve)* is determined by the space (flow area) between the movable needle and its seat, and by fuel-pump pressure. The needle's movement in relation to its seat is controlled by the float, which rises and drops with fuel level.

As fuel level drops, the float drops. This action opens the fuel inlet valve to allow more fuel to enter the bowl. When the fuel reaches a specified level, the float moves the needle to where it restricts fuel flow. Now, only enough fuel is admitted to the bowl to replace that being used. Any change in the fuel level causes a corresponding movement of the float: It opens or closes the fuel inlet valve to restore or maintain fuel level.

DESIGN FEATURES

Fuel Bowl—The fuel bowl, or float chamber, is a reservoir that supplies all fuel in the carburetor. Fuel level in the bowl is controlled by the float and fuel inlet valve. The pressure of fuel in the bowl is approximately the same as outside air (atmospheric pressure) because a vent connects the float bowl to the carburetor's air inlet passage.

Venting the fuel bowl to outside air prevents the fuel in it from being pressurized by the fuel pump. Fuel is pressurized up to the inlet valve, but once in the fuel bowl, it is at *vented pressure.*

The bowl also acts as a vapor separator. A vent in the bowl connects to the *inlet air horn.* Any vapors that may have been trapped in the fuel as it was pumped from the tank escapes through this vent, so only minor pressure is present in the bowl.

Many pre-emission-control carburetor models have a mechanical vent valve on the float bowl. At idle, or when the engine is stopped, this external vent releases fuel vapors into the engine compartment. The vent maintains constant near-atmospheric pressure in the fuel bowl. Restricted (dirty) air filters affect fuel metering with this design.

Connecting this vent to the inlet air horn guarantees the fuel bowl is vented to clean air from the air cleaner. Because the vent has the same pressure as the inlet air horn, air bleeds and idle system, a dirty air filter doesn't alter the air/fuel mixture. This holds true unless external venting is also used. Dirty air cleaners restrict airflow, creating a lower absolute pressure to the carburetor, which can cause a power loss.

The preferred location for a vent is near the center of the fuel bowl—high enough so fuel will not slosh into the air horn during abrupt maneuvers. Carburetors on most 1970-and-later cars have another vent connecting the bowl to a charcoal canister. When the engine is turned off, this canister collects escaping fuel vapor generated when engine heat causes gasoline in the fuel bowl to boil off.

The canister also captures fuel-tank vapor via plumbing from tank vents. Vapor collected in the charcoal canister is drawn

EFFECTS OF INLET SYSTEM CHANGES

Change	Effect
High float level	Raises fuel level in bowl. Speeds up main system start-up because less *depression*—reduction of pressure—is required at venturi to extract fuel from bowl. Start of flow from the bowl via the nozzle is sometimes called *pullover.* Increases fuel consumption. May cause fuel to spill through discharge nozzles and vent into carburetor air inlet on abrupt stops or turns to cause over-rich air/fuel mixture. Engine then runs erratically or stalls. This spillage affects low rpm emissions. Increases chances for the carburetor to *percolate* and *boil over.* This is a condition in which fuel is pushed by rising vapor bubbled out of the discharge from the main well when it is hot. Also, when the vehicle is parked on a hill or side slope, high float level may cause fuel spillage through the vent or main system.
Low float level	Lowers fuel level in bowl. Delays main system start-up because more vacuum is required at venturi to start fuel flow. This delay may cause flat spots or *holes* in power output. May expose main jets in hard maneuvering, causing *turn cutout*—misfire from lean air/fuel mixture. Can lessen maximum fuel flow at wide-open throttle—WOT.
Float assist spring too strong	Causes low fuel level. Inlet valve closes prematurely because of added force of spring.
Assist spring too weak	Causes high fuel level. Inlet valve closes after correct fuel level is reached because more float movement is required to compensate for weak spring.
High fuel pressure	Raises fuel level approximately 0.020-in. for each psi—pounds per square inch—fuel pressure increase. This factor varies with bouyancy, leverage, and needle-orifice size.
Low fuel pressure	Lowers fuel level
Larger inlet-valve seat	Raises fuel level.
Smaller inlet-valve	Lowers fuel level.
NOTE: Changing almost any part in fuel-inlet system requires resetting float to maintain correct fuel level.	

into the intake manifold when the engine is restarted. The vent to the vapor canister can be vacuum-operated, operated by vapor pressure from the bowl, or mechanically actuated at idle.

On some pre-emission-control carburetors, a bimetal-actuated vent releases fuel vapor to the atmosphere when it reaches a preset temperature. This prevents dust from entering the float bowl while driving, yet allows venting during *hot soaks*. This term refers to when an engine is stopped during hot weather or after it is fully warm.

Float—A float on a hinged lever opens the inlet valve so fuel enters the bowl when fuel drops below the reference height. The float then shuts off the valve when this height is reached. Floats on older carburetors are made from thin brass stampings soldered into an airtight assembly. These can develop leaks, fill with gasoline, and sink. Consequently, metering action is upset. The sunken float opens the fuel inlet valve, overfills the float bowl, and causes an over-rich air/fuel mixture.

Nowadays, most floats are made of closed cellular material that doesn't leak. But, they sometimes slowly absorb gasoline and lose their buoyancy. Once again, metering action is skewed. The more conditions change, the more they remain the same!

A *damping spring* is sometimes added to the float-arm pivot pin to minimize float vibration. Float vibration, caused by engine or vehicle vibration or bouncing, can cause wide fuel-level variations because a "lively" float allows the inlet valve to admit fuel when it is not needed. This float assist can be most helpful in preventing pronounced and unneeded *float drop*, which opens the inlet valve during short-turn/ acceleration maneuvers. Some carburetors may use a tiny spring inside the inlet valve itself instead of—or in addition to—a spring under the float. Some versions of the Corvair H carburetor had both damping methods. Such springs can be especially helpful in dirt-track, off-road and marine applications.

Every carburetor has a specified float setting. It is established to account for fuel pressure, float bouyancy, *seat size*—the inlet-needle orifice size—and the required fuel level in the fuel bowl. The correct float setting is the float location that closes the inlet valve at the required fuel level. This is

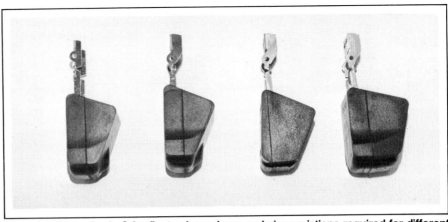

Closed-cellular plastic Q-jet floats show shape and size variations required for different fuel metering applications.

Arrow indicates clip connecting float to inlet needle for positive needle opening when float drops.

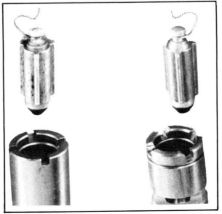

Two types of inlet valves. Non-windowed type on left flows more fuel in Q-jet. Note pull clips on needles.

referred to as the *mechanical setting* because a measuring instrument, not the fuel level, is used to determine its position.

The fuel level for a particular carburetor is established by the design and test engineers so the carburetor will operate without problems in all types of driving. The level is set so no fuel spills when a vehicle is parked or operated facing up, down or sideways on a hill with approximately a 30% grade.

For best operation in high-speed cornering, floats are *center-pivoted*, that is, the pivot axis is parallel to the vehicle's axles. For superior resistance to the effects of acceleration and braking, *side-hung* floats with pivots perpendicular to the axles have a slight advantage in float control and fuel-handling. Rochester Q-jets are the center-pivot type, as are the 4GCs. Two-barrels and Monojets are side-hung.

Fuel Inlet Valve—The rounded end of this valve rests against or—is connected to—the float-lever arm with a pull clip. The tapered end of the inlet valve closes against an inlet-valve seat as the float rises.

Some inlet valves are hollow and have a tiny damper spring and pin inside. This damper spring cushions the valve in its seat and protects it from road shocks and vehicle vibrations so its metering action isn't disrupted. The steel inlet valve in modern carburetors has a tapered seating surface tipped with *Viton*. This tip is extremely resistant to dirt and conforms to the seat for a good seal.

Inlet valves are supplied for various inlet-orifice sizes. The orifice size is not marked. To measure it, you must use plug gages or drill-bit shanks. Unless you have all number and fractional bit sizes in the 0.060—0.140-in. range, this measuring method won't be accurate.

Inlet-Valve Seat—The seat diameter and orifice length determine how much fuel flow occurs at a given fuel pressure. A smaller opening flows less fuel; a larger opening, more. Seat size is typically selected to handle a range of operating conditions. It should allow quick filling of the fuel bowl for acceleration after the car has been standing parked with a hot engine. And, the size should give minimum restriction for high fuel demands such as occur at wide-open-throttle (WOT) at high rpm.

Larger seats are also used for better purging of vapor from the fuel lines. A small-diameter seat controls hot fuel best because vapor pressure acts against less area when forcing the needle off its seat. Changing the seat size means changing the float level to balance the float against different fuel-pressure forces.

Fuel Filter—A fuel filter or screen may be found in the carburetor body or fuel bowl as part of the fuel-inlet system. The filter, which is between the fuel pump and the carburetor inlet valve, traps dirt that could cause inlet-valve seating problems.

This filter is installed with a pressure-relief spring behind it. If the filter clogs with dirt, the relief spring forces the filter off its base. Fuel then flows into the carburetor, even if this internal filter is clogged. In-line fuel filters are recommended and discussed on page 119.

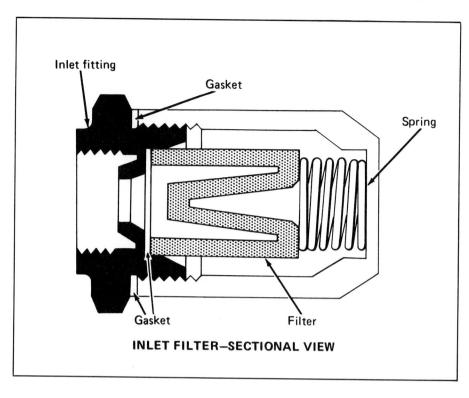

INLET FILTER—SECTIONAL VIEW

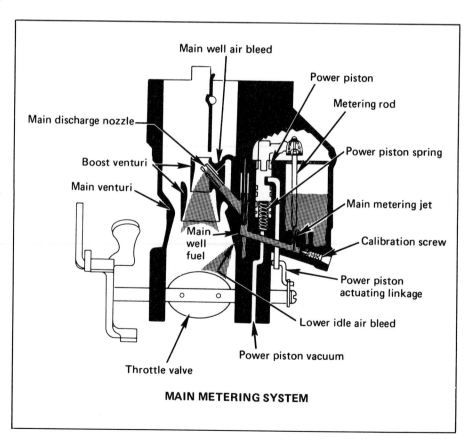

MAIN METERING SYSTEM

NEEDLE-TO-FLOAT CLIP

Why have a needle-to-float clip? If a vehicle or engine sits for a long time, deposits may accumulate in the fuel system. These deposits are heavy and sticky, and have a molasses-like consistency. In their first stages, they are known as *gum deposits*.

Cases have occurred where heavy gum concentration in poorly blended fuels formed enough harmful deposits to hang up inlet valves so they didn't follow the float's action. Most pump fuels today are carefully monitored according to standards set by organizations such as the American Society of Testing Materials (ASTM). Regardless, all fuels leave some deposits.

Be sure the clip is reinstalled so when the float moves, the inlet valve follows.

Bowl Stuffers & Baffles—Changes in speed, altitude or direction of a vehicle cause fuel to rush to the side, front or rear of the bowl. Abrupt changes may toss the fuel so that the main jets are uncovered, or fuel may even splash out of the internal vents into the air horn.

Three methods are used to counter these conditions. Baffles are sometimes inserted into the bowl to reduce the fuel-slosh area. Semi-solid air-horn gaskets are another fix. These may be solid in the area below vent openings, to serve as a baffle and prevent fuel from splashing from the vents. Bowl stuffers (inserts) are also used to reduce bowl volume and help control splashing fuel in the bowl.

MAIN SYSTEM

The main metering system supplies a good portion of the air/fuel mixture to the engine for cruising, acceleration and medium-throttle driving. The system primarily consists of fuel that flows through the main jet(s). Different systems of the carburetor come into play under other driving conditions.

OPERATION

During the time engine speed or airflow is increased by throttle demand to the point where the main metering system begins to operate, some fuel is fed by the idle and accelerator pump systems. Under conditions of high load, when the engine must produce full power, added fuel comes from the power system. These are examined in detail later in this chapter.

Throttle—Many people believe the throttle controls the volume of air/fuel mixture pumped into the engine. This is not the case. Piston displacement never changes, so the *volume* of air pulled into the engine is constant for any given speed. The mixture volume entering the engine is calibrated to meet the need.

The throttle controls the *density,* or mass flow, of the air pumped into the engine by the action of the pistons. Lowest density of charge is available at idle and highest density is at WOT. A dense charge has more air mass, so higher compression and combustion pressures can be developed for higher power output.

Simply stated, the piston displaces the same volume during each intake stroke.

When the throttle is nearly closed, the piston pulls a thinner charge—less dense—because of throttle-blade restriction. With more throttle opening, the airflow is less restricted and the cylinder can inhale volume nearer its displacement. Thus, the throttle controls engine speed and power output by varying the *charge density* to the engine.

Venturi—This is the heart of the carburetor and the key to its simplicity. Understanding the venturi is fundamental to grasping the inner workings of the main system. So, let's closely examine it before progressing further.

The venturi and its principle of operation are named after G.B. Venturi, an Italian physicist (1746-1822). He discovered that when air flows through a constricted tube, flow is fastest and the pressure lowest at the point of maximum constriction. In your car's engine, a partial vacuum is created in each cylinder by the down-stroke of the piston. Because atmospheric pressure is nominally 14.7 psi, air rushes through the carburetor and into the cylinder to fill this partial vacuum. On its way to the cylinders, the air passes through the venturi.

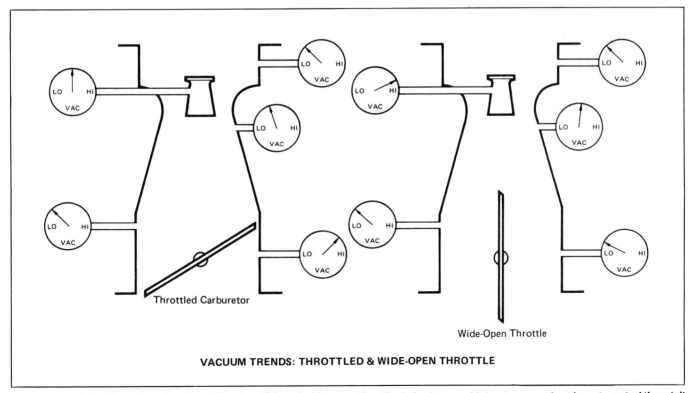

Throttled Carburetor

Wide-Open Throttle

VACUUM TRENDS: THROTTLED & WIDE-OPEN THROTTLE

Vacuum signals inside carburetor air section at partial- and wide-open throttle: In both cases highest vacuum is at boost-venturi throat; it signals main system.

Physically, the venturi is a smooth-surfaced restriction in the path of the incoming air. It constricts the inrushing air column, then allows it to widen back to the throttle-bore diameter. Air is flowing in with a certain pressure. To get through the constricted area, it must speed up, which reduces the pressure inside the venturi. A gentle diverging section is used to recover as much of the pressure as possible. This section starts at the smallest area of the venturi and continues to the lower edge of the tapered section.

The venturi controls the main metering system by governing the fuel discharged into it at the point of lowest pressure, or greatest vacuum. This minimum-pressure point supplies a signal to the main metering system. This pressure-drop signal travels to the main and power systems via an *asperator channel*. This channel is plumbed from the point of greatest depression in the boost venturi on an upward angle to the vertical fuel pickup channel. The channel, in turn, gets fuel from the main and power systems.

The pressure-drop or *vacuum-signal* requirement is designed and flow-tested to supply the required signal at the discharge nozzle in the venturi. Because the fuel bowl is kept at near-atmospheric pressure by the vent system, fuel flows through the main jet(s) and into the low-pressure or vacuum area in the venturi.

Pressure drop at the venturi varies with engine speed and throttle position; the vacuum increases with engine rpm. Consequently, WOT and peak rpm cause greatest airflow and highest pressure difference between the fuel bowl and a discharge nozzle in the venturi. Highest fuel flow occurs at these conditions.

The pressure drop in the venturi also depends on its size. A small venturi has a higher pressure drop at any given rpm and throttle opening than a large one.

Main Air Bleed—Once main-system flow is started, fuel is metered, or measured, through a main jet in the fuel bowl. From the main jet, fuel passes into a main well. As fuel passes up through the main well, air from a main air (high speed) bleed is added to pre-atomize or emulsify the fuel into a light, frothy air/fuel mixture. This mixture issues from the discharge nozzle into the air steam flowing through the venturi. The discharge nozzle is often located in a small boost venturi centered in the main venturi.

Top view of primary Q-jet bore shows two booster venturis in a cluster or stack to increase signal for improved metering at low airflows. Some carburetors have single boost venturi.

There are two main reasons why liquid fuel is converted into an air/fuel emulsion. First, the emulsion vaporizes much easier when it is discharged into the air flowing through the venturi. Second, it has a lower viscosity than liquid fuel and responds faster to any change in the venturi vacuum—signal from the venturi applied through the discharge nozzle. It will start to flow more quickly than pure liquid fuel.

The strong signal from the discharge nozzle is bled off (reduced) by the main air bleed so there is a less effective pressure difference to cause fuel flow. The mixture will become leaner as the size of the bleed is increased. Decreasing the bleed size increases the pressure drop across the main jet, so more fuel is pulled through the main system, resulting in a richer mixture.

Changes to the main air bleed affect the main metering system throughout its operating range. RPD establishes the main-air-bleed size for each carburetor to work correctly for its application. The main air bleeds on RPD carburetors are not removable as are those on some other carburetors.

Changes in main-air-bleed sizes are rarely necessary or advisable in field service. This is a valuable calibration aid in original calibration work. Calibration changes are easy to make by changing main metering jets.

The main air bleed also acts as an anti-siphon, or siphon breaker, so that fuel does not continue to dribble into the venturi after airflow is reduced or stopped.

DESIGN FEATURES

Throttle—The throttle shaft is offset slightly—about 0.02 in. on primaries and about 0.06 in. on secondaries—in the throttle bore so that one side of the throttle has a larger area. This design causes self-closing. It improves idle-return consistency because sizable closing force is generated when the manifold vacuum is high—as at idle. Also, it is a safety measure to guard against overspeeding the engine if it is started without installing the linkage or throttle-return spring.

Primary throttles are seldom closed against the throttle bore; instead, they are factory-set against a stop for closed-throttle airflow specified for that particular carburetor model. In most cases, this stop is the idle-speed stop screw.

Venturi—The most efficient venturi—one that creates the maximum pressure drop with the minimum flow loss—requires a 20° entry angle and a diverging section with a 7—11° angle at its end. Designers try to keep as close as possible to these ideal venturi entry and exit angles.

Although the theoretical point of lowest pressure and of highest velocity would be expected at the minimum diameter of the venturi, friction causes this point to be at about 0.03-in. below it.

A venturi generates a stronger metering signal than a straight tube and has minimal loss of air pressure due to friction. This is because its trailing edge conforms to the normal air stream and recovers most of the pressure drop so a greater mass of air is available to the engine. This pressure recovery would be much less if a flat-plate orifice were used. The venturi is an efficient air-measuring device because of the high signal levels it produces with minimal pressure loss.

If engine displacement is constant, a larger-venturi carburetor will require a higher rpm to bring the main system into operation than one with a smaller venturi. Venturi size also controls the quantity of air available at WOT. If too small, top-end HP will be reduced, even though the carburetor supplies efficient air/fuel vaporization at all speeds.

Automobile manufacturers typically compromise by using carburetors with air-flow capacities smaller than the maximum. They choose these types of carburetors, even though you—a performance enthusiast—may feel cheated. Efficient fuel vaporization promotes good distribution and improves smooth running and economy at around-town and cruising speeds. These are certainly characteristics most manufactures want to promote!

Manufacturers who offer more performance usually supply a carburetor with a secondary system. This design retains the advantages of a small primary venturi with the capability of higher airflow for top-end power. The Q-jet is an excellent example of this type.

Boost Venturis—These aid fuel distribution because the ring of air flowing between the two venturis directs the charge to the center of the air stream. This effect helps keep wet air/fuel mixture off the carburetor wall below the venturi—more air/fuel mixture reaches the manifold for improved vaporization. Tabs, bars and wings are sometimes added to cause correct directional effects for good cylinder-to-cylinder distribution with a particular manifold.

Boost venturis permit the use of a shorter main venturi, so a carburetor can be compact. Carburetor designers could achieve essentially the same results with a long venturi as with one or more boost venturis stacked in the main venturi. But it's tough to build a carburetor with ideal entry and exit angles, and simultaneously reduce its diameter for an adequate signal. Some RPD carburetors use two boost venturis to get a signal for main-system operation.

Main Jets—These metering orifices control fuel flow into the metering system. They are rated in flow capacity and are removable for calibrating a carburetor. It is commonly thought that main jets are selected solely by trial and error, but this is not the case. For a given venturi size, a small range of main-jet sizes will cover all conditions. The final selection of the correct jet for an application is done by testing because design and operating variations—climate, altitude and temperature—affect jet-size requirements.

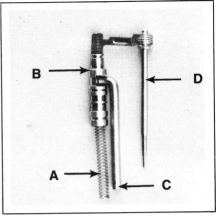

Monojet power metering-rod assembly shows power spring (A), power piston (B), actuating rod (C) and metering rod (D) that centers in main metering jet.

Screw-in power valve is actuated by power piston responding to manifold vacuum. As vacuum drops, power-piston spring opens power valve. Power system fuel supplements that supplied by main jets.

Cutaway of Q-jet primary bore showing where fuel discharges from center of smallest booster venturi. Arrows indicate fuel nozzle openings.

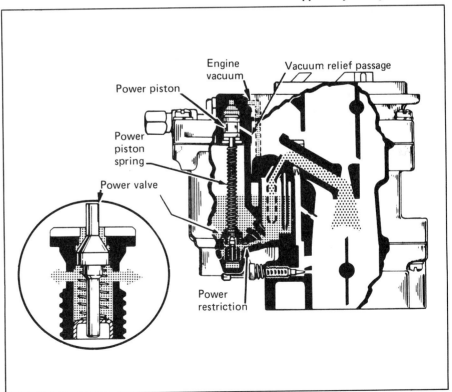

There is a basic misconception about jets: Size alone determines their flow characteristics. This is incorrect because the shape of the jet entry and exit, length of orifice and finish affect flow. RPD checks each jet on a flow tester and grades it according to its flow. This checking is a production-line calibration method used to assure the customer uniform air/fuel ratios.

The tolerance range used during production for each size explains why a 69 jet (number equals orifice size in thousandths: 0.069 in.) found in a carburetor may not seem to give a richer mixture than a 68. If the 68 is on the high side of its allowable tolerance and the 69 is on the low-side tolerance limit, the two jets will flow almost the same amount of fuel. The tolerance range for a 68 is from 382 to 398 cubic centimeters (cc) per minute with a given test fuel. The 69 ranges from 394 to 410cc.

Service jets will have a mean (average) flow rate. For instance, a 69 jet, will have approximately 402cc flow. For fine tuning, it is best to use sets of jets from the same origin. I prefer RPD service jets, because these have a mean flow characteristic rather than a high or low tolerance. Don't use one jet from the carburetor, another from your dealer, and still others from a specialty

shop. There are enough variables to tuning without introducing unneeded ones!

A plus for tuners is that RPD supplies only the mean flow jets for service replacements. Therefore, when you select a 72 jet to replace a 71, you can be sure it will flow more fuel. Jets used in RPD carburetors supplied to auto manufacturers may vary due to flow differences, but the carburetor has been checked on a *flow stand* to guarantee its mixture characteristics fall within allowable tolerance. This allowable production tolerance means that a 71 jet in a new carburetor or one supplied on your automobile, truck or marine engine could flow about the same as a 70 or a 72.

Drilling out jets to change their size is never recommended for carburetors used in normal passenger-car driving. It is a more accepted practice in high-performance applications. Drilling can destroy the jet's entry and exit features and introduce a swirl pattern.

Main Air Bleed—All Rochester carburetors are equipped with built-in fixed-dimension air bleeds. If you're accustomed to removable air bleeds found in many other makes, this may seem a limitation. More comment on modifying air bleeds is found on page 136.

POWER SYSTEM

When the engine is required to produce power in excess of normal driving requirements, the carburetor must supply a richer mixture. Additional fuel is supplied by the power system, under control of manifold vacuum. Manifold vacuum accurately indicates the load on the engine. Vacuum is usually strongest at idle. As load increases, the throttle valve must be opened wider to maintain a given speed, thereby offering less restriction to air entering the intake manifold, reducing manifold vacuum.

Power Valve—A vacuum passage in the carburetor applies manifold vacuum to a power-valve piston. At idle or cruising, manifold vacuum acting against a spring holds the valve closed. As high power demands load the engine, and manifold vacuum drops below a certain point, the power-valve spring overcomes manifold vacuum and starts to open the power valve.

Fuel flows through the power valve, then through a restriction, to join fuel already coming from the main metering system. Consider the power valve a switch that turns on extra fuel under high engine loads.

Race-car engines often have large manifold-vacuum fluctuations at idle and low speeds. Vacuum can be low enough to cause the power valve to begin feeding additional fuel. The valve must be modified to not open and close in response to these fluctuations, which are caused by valve timing instead of throttle position or engine load. The power valve must still function, but at a lower vacuum than occurs at idle, as explained in the High Performance Carburetion chapter.

Design features of some power systems are in photos on pages 11 and 12.

ACCELERATOR PUMP SYSTEM

The accelerator pump:

• Makes up for fuel that condenses on manifold surfaces when the throttle is opened suddenly.

• Acts as a mechanical injection system to supply fuel before the main system starts.

OPERATION

When the throttle is opened quickly, intake-manifold vacuum instantly drops, increasing manifold pressure toward atmospheric. A high manifold vacuum keeps

2G power system uses separate power valve (at bottom) with vacuum-piston (arrow) actuated by manifold vacuum. Spring on piston stem overcomes lowered manifold vacuum and piston end pushes center stem of valve for fuel flow.

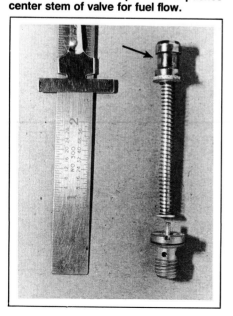

Q-jet power piston assembly has two metering rods: one for each primary bore and main jet.

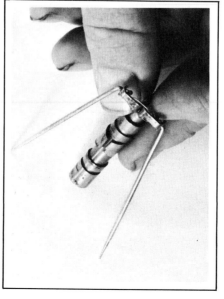

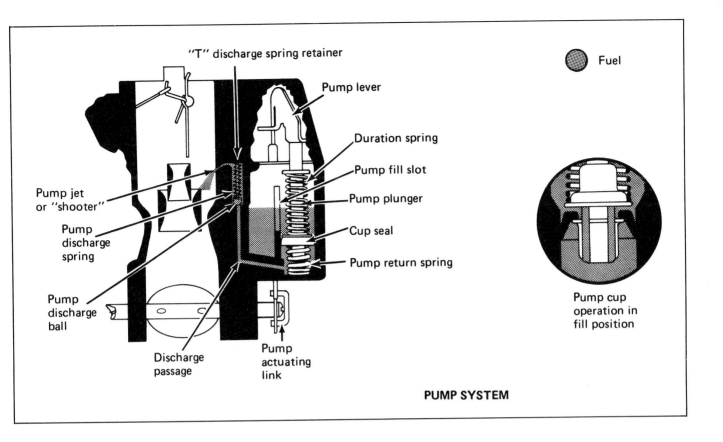

"T" discharge spring retainer

Fuel

Pump lever

Duration spring

Pump fill slot

Pump plunger

Cup seal

Pump return spring

Pump jet or "shooter"

Pump discharge spring

Pump discharge ball

Discharge passage

Pump actuating link

Pump cup operation in fill position

PUMP SYSTEM

the mixture vaporized. As pressure approaches atmospheric, fuel condenses into puddles and wet spots in the intake manifold. The mixture available to the cylinders is instantly leaned out and hesitation or stumbling occurs unless fuel is immediately added.

This additional fuel is especially important in big-port manifolds or manifolds with large plenum areas because these have more surface area for fuel to condense on. It is also needed before the engine is thoroughly warmed up.

Another function of the accelerator pump is to inject fuel during abrupt throttle transitions when the idle system supply is passing to the main system. The accelerator pump delivers fuel until the main system can begin operating.

Accelerator-pump operation is examined in the Q-jet Design, Q-jet Service, and High Performance Carburetion chapters.

DESIGN FEATURES
Pump Inlet Valves—RPD designs of the mid-'60s allow fuel to enter the slotted pump well. It flows past a check ball in the plunger head and around the pump plunger. Downward motion of the plunger seats the check ball so fuel is forced through the pump-system discharge channel. The check ball also serves as a vapor vent to relieve any vapor pressure that might form in the pump well.

Some RPD models have an inlet check ball at the bottom of the pump well. This ball is in a pump-inlet passage supplied from the bowl. The passage may be protected by an inlet screen.

The floating-cup design is another type of pump inlet valve. Accelerator pumps used on most current RPD carburetors fill through the center of the pump cup. The cup is fitted onto the plastic pump-plunger body with some vertical clearance.

During the delivery stroke, the cup is forced up against the plastic pump-piston face, sealing off the fill hole positioned there. When returning upward, the cup drops a few thousandths of an inch from the piston face, and fuel enters through the center of the cup. Fuel fills the well through this hole so all is ready for the next shot.

Q-jet pump cutaway showing pump cup at bottom of stroke with return spring compressed. Duration spring is on top of cup.

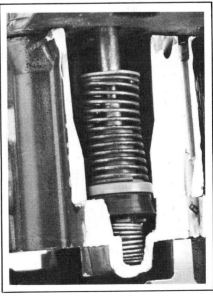

The pump-shot and refill are designed to allow full-capacity shots at intervals of two to three seconds.

Pump capacity depends on piston size and stroke. Measure the capacity as follows: Funnel the pumped fuel into a graduated cc beaker (burette). The fuel level should be maintained by wiring an electric fuel pump to supply fuel at operating pressure. Now, consecutively activate the throttle from closed to full-open 10 times. Move the throttle steadily and hold it in the full-open position approximately one second each time. This allows the fuel shot to be expelled to the burette.

At the end of 10 strokes, the cc of collected fuel is the pump capacity. Small carburetor pumps will pump 3—5cc. Larger carburetors will pump 15—20cc in 10 strokes.

Methods to increase Q-jet carburetor pump capacity to nearly 50cc are discussed on page 128.

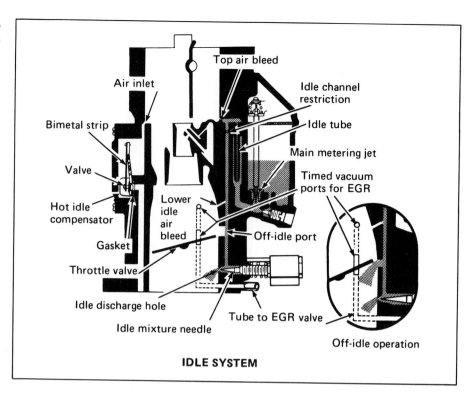

IDLE SYSTEM

IDLE SYSTEM

For many years carburetors required a richer mixture at idle than at part-throttle operation. Unless the idle mixture was richer, slow and irregular combustion would result because residual exhaust gases diluted the mixture. Since the early '80s, idle systems are set nearer to a 14.7:1 air/fuel ratio to match emission hardware needed to meet clean-air standards.

OPERATION

The idle system supplies fuel at idle and low speeds, and should keep the engine running when accessories are on. It also must keep the engine running against the load caused by placing an automatic transmission in gear.

At idle and low speeds, not enough air is drawn through the venturi to cause the main metering system to operate. Intake-manifold vacuum is high because airflow is restricted by the nearly closed throttle valve. This high vacuum supplies the pressure differential for idle-system operation.

When the throttle is closed—or nearly so—reduced pressure between throttle and intake manifold draws the air/fuel mixture through the curb-idle port in the bore *below* the throttle plate. When the throttle first starts to open, manifold vacuum is still high and additional mixture is drawn through the

off-idle port as it is uncovered by the opening throttle. The amount of flow through the idle system depends on channel restriction, size of the idle-discharge ports, and idle-mixture screw setting.

Off-Idle System—Most RPD carburetors have an intermediate idle system that discharges fuel through a hole below the venturi. This hole is drilled into the vertical feed passage of the idle-fuel-supply passages from the bowl. It serves two purposes.

First, it bleeds air into the vertical fuel-supply channel when the idle system is operating. This air provides better air/fuel mixture before it enters the carburetor bore through the curb-idle and off-idle holes or slots.

Second, because of its location between the venturi and the throttle blade, the bleed hole is subjected to low pressure, or vacuum, when the throttle blade is opened past the off-idle slot and to a point near the lower idle-air bleed hole. At this point, the lower idle-air bleed becomes a feed to supply fuel to cover up any lag between the

off-idle fuel system and the start of the main system.

When this bleed/feed has a strong signal—when throttle blade is near the bleed/feed hole—the idle/off-idle holes/slots will supply little, if any, fuel. This lower idle-air bleed/feed gets fuel from the idle system via idle-metering orifices connected to the bowl.

Supplemental fuel from the lower idle-air bleed/feed diminishes as the throttle opens to a point where the main system becomes self-sustaining in its flow.

A larger bleed/feed hole can supply more fuel when the throttle blade passes it. Thus, it can be effective in covering up any "lean hole" or "sag" after the throttle passes the off-idle holes/slots and before the main system starts to flow. When it acts as a bleed during idle/off-idle operation, the larger hole—bleed—will lean near-closed-throttle mixtures.

Flow from the idle system tapers off as the main system starts to discharge fuel. Both systems are designed to achieve a gradual transition from idle to cruising speeds when carburetor capacity is correctly matched to engine displacement.

In normal driving, flow swings quickly back and forth between idle and main operation as the vehicle is accelerated, slowed by closing the throttle, idled at stop, and then reaccelerated.

DESIGN FEATURES

Idle Air Top Bleed—Besides the feed/bleed combination just described, there are one or two fixed air bleeds at or near the top of the idle-fuel-feed channels. Increasing their bleed size reduces pressure drop across the bleed, decreasing the amount of fuel brought from the idle well. Increasing idle-air-bleed size leans the mixture, even if the idle-feed restriction is left constant. Conversely, decreasing the size of the idle-air-bleed increases the amount of pressure drop and richens the idle mixture.

Auxiliary Air Bleeds—Auxiliary air bleeds are sometimes used in the idle system. Although these usually add air to the idle system downstream from the traditional idle-air-bleed, they act in parallel with the idle-air-bleed.

Idle-Speed Setting—Before emission-control requirements became important, idle setting was the slowest speed that would keep the engine running smoothly. Emission requirements have made higher idle speeds necessary. Higher idle speed reduces some of the exhaust-gas dilution that occurs at lower idle speed, so a leaner idle mixture can be used without misfiring.

Cars with engines designed to pass emission requirements, i.e., production vehicles, are typically set for leanest best idle at a specified rpm. Then, the idle is reduced by leaning the mixture still further. The manufacturers have carefully correlated this leanest best idle and subsequent idle drop-off to meet a legislated minimum carbon monoxide percentage (CO%).

The idle on older non-emission-controlled cars and race cars is typically set for the desired idle rpm and best manifold vacuum. This is not a minimum emission setting.

Idle Limiter—An idle-limiter cap limits idle-mixture screw adjustment to less than 1/2 turn. This limiter is installed after the desired idle mixture has been factory-set, to prevent easy tampering. The limiter is constructed such that it will be destroyed if you remove it. Consequently, it indicates the idle mixture has been readjusted, perhaps out of emission specifications.

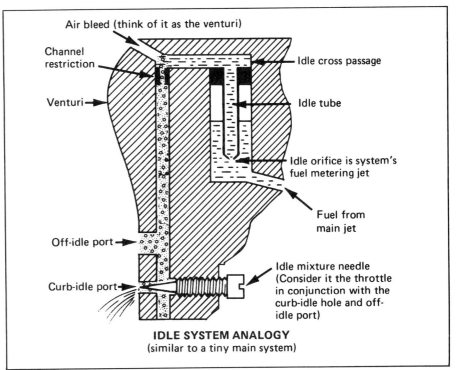

IDLE SYSTEM ANALOGY
(similar to a tiny main system)

Idle system operates like main system, only on smaller scale.

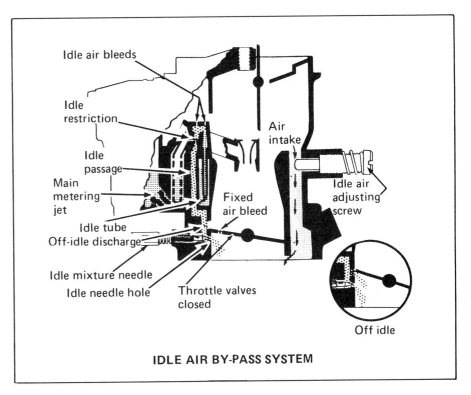

IDLE AIR BY-PASS SYSTEM

Idle-Air Bypass—Some RPD two- and four-barrel carburetor models have this system. It allows throttle valves to be almost closed at idle. Gum and carbon deposits around throttle valves of conventional systems often disrupt engine idle, but RPD's design ensures a consistent idle, even with a dirty carburetor.

Idle air from the carburetor bore *above* the venturi bypasses the closed throttle valves through a separate air channel and enters the carburetor bore just *below* the throttle valves. The amount of air to the engine is regulated by an idle-air-adjustment screw in the idle-air bypass channel. This screw is mounted at the rear of the float-bowl casting. Turning the screw in (clockwise) decreases idle speed; backing it out (counterclockwise) increases it.

This system is a superior way to reduce *nozzle drip* at idle. Air rushing by the venturi often causes this problem, and with the bypass system, much of the idle air sidesteps the venturi.

To obtain sufficient idle air for adequate idle speed, a fixed idle-air supply is necessary with the adjustable air or bypass supply. The fixed idle-air path in a two-bore carburetor is provided by a hole in each throttle valve. These fixed idle-air holes maintain a constant idle-air flow for part of the idle-air requirements, while the idle-air-adjustment screw regulates the remainder of the idle air. Thus, idle speed is adjusted by the idle-air-adjustment screw. The throttle plates are not moved from their factory-set positions.

Fixed Idle-Air Bypass—A few Q-jet models have idle-air channels from the air horn to a point below the primary throttle valves. Extra idle air coming through these channels allows the throttle valves to be more closed at idle, reducing the signal applied to the main fuel nozzles by the efficient venturi cluster. The bypass eliminates nozzle drip at idle, an especially annoying problem on large-displacement engines operating at the high idle speeds required for emission control.

Idle-Air Compensator—This is used on some RPD two-barrels to offset enriching effects caused by excessive fuel vapor, which is created by fuel percolation during extremely hot-engine operation.

The compensator is a thermostatically controlled valve, usually mounted above the main venturi or at the rear of the float bowl. The valve closes an air channel leading from above the carburetor venturi to a point below the throttle valves. The valve is operated by a temperature-sensitive bimetal strip.

During extremely hot engine operation, excessive fuel vapor enters the manifold and causes too rich a mixture. The engine idles poorly or stalls. At a precalibrated temperature, when extra air is needed to offset the extra fuel vapor, the bimetal strip bends and unseats the valve. This uncovers the compensating air channel that runs from the carburetor venturi to a point below the throttle valves. Then, just enough air is added to the mixture to offset the richness and maintain a smooth idle. When the engine cools and the extra air is not needed, the bimetal strip closes the valve and operation returns to normal mixtures.

To make correct idle adjustments, the valve should always be closed when setting idle speed and mixture.

Curb-idle discharge is through hole at (A). Off-idle slot (B) just behind throttle plate is barely visible. Fixed idle-air by-pass orifice is at (C).

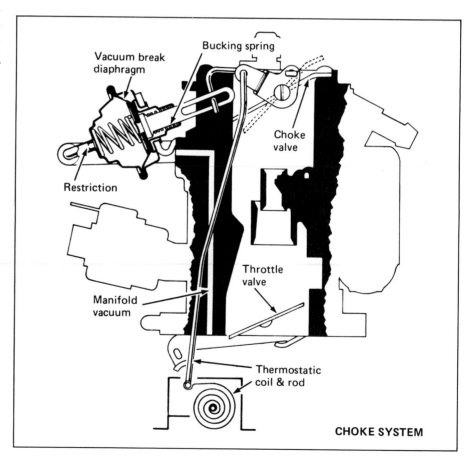

CHOKE SYSTEM

Adjustable Off-Idle Air Bleed—Some emission-control carburetors include an adjustable off-idle (AOI) air-bleed system. A separate air channel bleeds air past an adjustment screw (needle) into the idle system. Factory adjustment of the screw establishes precise off-idle air/fuel mixture ratios to meet emission-control requirements.

This system is not common. It is simply an idle-system air bleed with an adjustable needle used to alter bleed size.

Idle-Channel Restrictions—Some idle systems have a calibrated restriction in the idle channel. This secondary metering for the idle system affects transition metering from 25 to 40 mph, depending on the carburetor.

Curb-Idle Port Size—Emission-control carburetors may have smaller idle-mixture-screw-discharge holes than those used on pre-emission carburetors. This was done to eliminate over-richening caused by backing out the mixture-adjustment screws. As an example, 1970-and-earlier Q-jets have 0.095-in.-diameter curb-idle discharge holes. Later models have 0.080-in. or smaller holes. In some instances, no matter how far you back the idle-mixture screws out, it is not possible to create a richer mixture.

Off-Idle Discharge Ports—These can be either one or more holes or a slot. Either type provides correct air/fuel mixtures and satisfactory operation. Slots are usually used because they are less expensive to manufacture.

These function like the air-bleed/fuel-feed orifice previously described. When the port is above the blade, it is an air bleed; when it is below, it feeds fuel.

CHOKE SYSTEM

Although it looks like a simple valve atop the air horn, the choke system is often one of the most complex in the carburetor. It controls the richer mixture required to start and operate a cold engine. That may sound easy, but cold-starting conditions are not ideal.

OPERATION

Cranking speeds for a cold engine are often less than 100 rpm, so little manifold vacuum is created to operate the idle system. A closed choke valve causes a vacuum below it and fuel is pulled from the idle and

main metering systems during cranking; at times fuel is even extracted from air bleed holes. All this surplus fuel produces an extremely rich mixture of approximately 5:1 air to fuel. This proportion is needed because very little fuel reaches the cylinders as vapor during cold starts.

Cold-starting inhibits fuel vaporization. Minimal manifold vacuum is also an impediment. The manifold is cold, so most of the fuel puddles around the manifold surfaces. The fuel is cold and not sufficiently vaporized; so it is not as volatile. Finally, liquid fuel cannot be evenly distributed to the cylinders and will not burn correctly. Once the engine does start, the mixture is

leaned-out somewhat because increased air flow partially opens the choke.

In an automatic choke, a *vacuum break diaphragm* pulls the choke valve to a pre-set opening once the engine starts. This partially leans the starting mixture. With the choke so positioned, it causes a 20—50% richer-than-normal mixture. As the engine warms and the bimetal choke thermostat weakens, the choke opens fully and mixture is stabilized. To reduce emissions, automatic chokes are usually calibrated to remain open the first 0.8 to 1.5 miles of city driving. Intake runner temperature must reach 55—75F (13—24C) before choke action is reduced.

Many carbs typically use integral choke. (A) indicates connection for exhaust-heat tube, (B) points to choke index.

Installation showing stat coil mounted on intake manifold over exhaust-heat crossover passage.

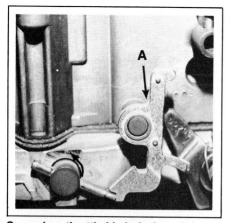

Secondary throttle-blade lockout (A) prevents secondary operation until choke link is pulled away by choke opening. (Some of choke mechanism removed for clarity.)

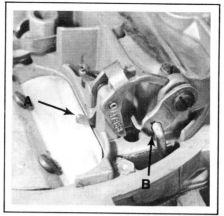

Some Q-jets use air-valve lockout instead of locking out secondary throttles. At left, choke is partially on, (A) indicates lockout lever; (B) is connecting link to fast-idle cam and thermostatic coil. At right, choke is fully off and air-valve lockout has retracted to permit secondary operation.

DESIGN FEATURES

Automatic Chokes—These are usually of two types: *integral* and *divorced*. The integral type uses a tube to direct heated air from the exhaust manifold or crossover to a bimetal spring inside a housing on the carburetor. The divorced type uses a bimetal spring mounted directly on the intake manifold or in a pocket in the exhaust-heat passage of the intake manifold. A mechanical linkage from the spring operates the choke on the carburetor.

Integral Choke—In this arrangement, the thermostatic coil is on the carburetor. Hot air is applied to the coil from a stove around the exhaust manifold. An integral choke heated by hot air can close when the engine is still hot, even though a rich mixture and fast idle are not needed. This is caused by cold underhood air, and results in too rich a mixture and starting problems after a partial warmup.

When the heat source is engine water, the choke is called a *hot-water choke*. It also can suffer the same restart problems described above.

You adjust an integral choke by changing the setting of the *choke index*. Usually, the thermostatic coil is set at the factory so the choke will close when the bimetal coil reaches 75F (24C). Many emission-calibrated carburetors have riveted choke covers, so resetting these takes some tenacity and engineering.

If less choke is needed at this temperature, then the choke index is moved one mark to decrease bimetal tension. If more choke is needed, the index is moved in the opposite direction. On most 1970-and-older units, an arrow on the choke housing shows which direction leans or richens the choke mixture. A choke-mixture change rarely requires moving more than one mark from the factory setting.

Divorced Choke—The divorced choke is operated by a thermostatic coil on the intake manifold and connected to the choke blade by a rod. You adjust a divorced choke by changing the rod length. Shortening it applies more closing force to the choke so it closes at a higher ambient temperature and stays shut longer.

Fast Idle—An engine has higher frictional forces to overcome when warming up. Consequently, higher idle speed is required to prevent stalling. A fast-idle screw on the throttle contacts a cam linked to the choke. Cam rotation holds the throttle valve open farther, to increase idle rpm. When the choke valve is fully open, the fast-idle cam disengages from the fast-idle screw and idle returns to normal.

Secondary Lockouts—When an engine is warming up, the secondary throttles shouldn't function. The sudden burst of air to the cylinders via cold manifold runners causes hesitation or backfire. To prevent these, *secondary lockouts* are connected

with the choke. When the choke is on, the lockouts keep the secondary throttles from working. After engine warmup, the choke comes off, allowing the secondaries to open.

Vacuum Break—This term originates from the action of closing a vacuum passage, thereby breaking the vacuum. Carburetors with a choke thermostat housing on their body have a manifold-vacuum passage to draw exhaust-heated air across the bimetal strip that controls choke opening. This vacuum also operates a piston or diaphragm that forces the choke slightly open once the engine starts—just far enough to allow the engine to run without loading or stalling.

Choke-valve position is determined by the torque of the thermostatic coil balanced against vacuum-piston or diaphragm pull, and air velocity acting on the choke valve. Because the piston or diaphragm operates from manifold vacuum, any heavy acceleration removes the vacuum, slightly closing the choke to richen the mixture for acceleration.

On models with the piston arrangement, once the choke opens—because of air velocity or heating of the thermostatic coil—the piston covers the vacuum port and breaks the vacuum. The term vacuum break is still applied to any vacuum-operated choke mechanism, including the diaphragm types, even though these don't include a vacuum break, as such.

Unloader—This counters starting problems caused because the engine is hot but the choke thermostat coil is cold. Because the choke coil and housing are external engine parts, they cool before the engine's internal parts. If an engine has been shut off for about an hour, this temperature difference can be substantial.

Thus, if the choke coil has cooled, it signals for full choke (maximum enrichment) when the engine is restarted, even though the engine is still warm. The choke sends a full cold-start supply of fuel that readily vaporizes in the combustion chambers. But sometimes this mixture is too rich for the warm engine to fire.

To start the engine, apply full throttle: The unloader linkage will open the choke blade and admit fresh air to the induction system. The mixture is now combustible. The unloader is actually a tang on the throttle lever that contacts the fast-idle cam and opens the choke plate enough to clear, or unload, excess fuel from the manifold.

The unloader aids in starting a flooded engine, regardless of the cause. Many people are not familiar with its function and continue to crank a flooded engine without completely opening the throttle. Some even add to restart problems by pumping the accelerator. Also, if the throttle is fully opened, reduced manifold vacuum on the choke vacuum-break piston or diaphragm will tend to close the choke and cause an excessively rich mixture when it is not required. The unloader ensures that the choke valve will be mechanically forced open so air can flow through the carburetor and purge the over-rich mixture.

Choke stat cover removed to show bimetal in cover. Tab or loop in end of bimetal attaches to choke tang (arrow).

WHAT'S BIMETAL?

The bimetal component in an automatic choke consists of two different metals bonded into a strip and formed into a coil. Because the metals have unequal thermal expansion characteristics, the coil unwinds when warmed and winds when cooled.

The outer (free) end of the coil attached to the choke linkage holds the choke plate closed—or loads it to close the plate when the throttle is opened—until the bimetal is warmed. Warming can be by exhaust-warmed air, jacket water, heating of the manifold from engine operation, or an electric heating element. Bimetal-temperature response depends on the metals used in the strip. Most automotive choke bimetals are calibrated to close the choke at 75F (24C).

SECONDARY THROTTLE

For many years U.S.-made carburetors were single-staged. The V8 had a two-barrel, but it was really only a single casting with two side-by-side one-barrels—one for each level of the traditional two-level cross-H manifold. All past and current Rochester two-barrels are single-staged: The throttles operate together on one shaft.

In the late '40s increased emphasis was placed on vehicle performance. Single-stage carburetors became bigger and bigger—and so did performance problems. The large venturis caused main systems to start flowing at higher rpm: Under some conditions the idle system could be made to mask main system late entry. But, because the idle system was controlled by manifold vacuum, a power deficiency resulted at low manifold vacuum and low airflow levels. In addition, weak venturi velocities at low rpm caused poor fuel vaporization and air/fuel-mixture quality. Cylinder-to-cylinder distribution problems and erratic operation resulted.

Simply stated, the capacity of one- and two-barrel carburetors was too limited to satisfy all driving and performance require-ments. The answer was obvious: Use a *staged* carburetor. This design expands the metering range and uses responsive primary venturis small enough to get the main systems flowing at low rpm and promote good vaporization, while having capacity for high-rpm operation. Staged four-barrel carburetors became popular in the late '50s and remained so through the mid-'80s. Staged two-barrels were used on small displacement/low HP engines in the early '80s to meet emission and fuel economy standards.

The secondary system in a staged carburetor is simply another carburetor in parallel with the primary. Secondaries always have a main metering system, and occasionally an idle system. Q-jets don't have secondary idle systems.

On carburetors with a secondary fuel bowl, the secondary idle system keeps the fuel level from rising in the bowl if the needle and seat leak or open due to float bounce. The idle system is sometimes used to purge the secondaries—supply fresh fuel—to prevent varnish caused by a driver not using them. RPD generally does not include an idle system in the secondary side.

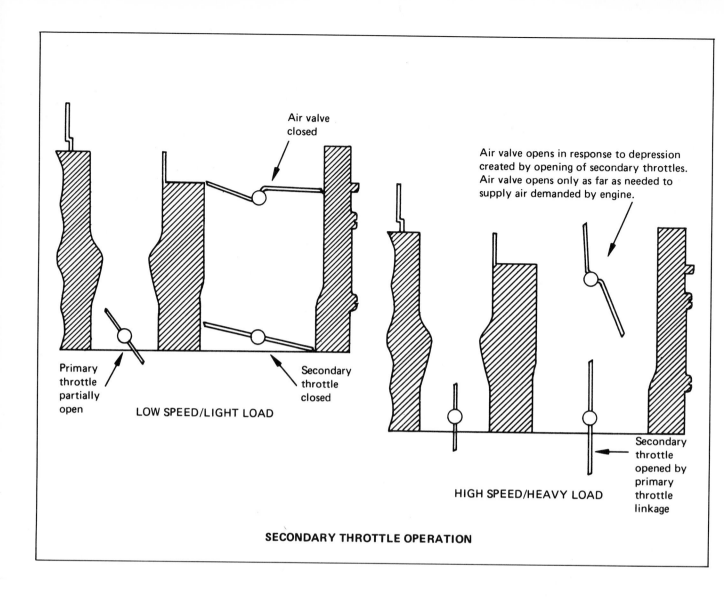

Air valve closed

Air valve opens in response to depression created by opening of secondary throttles. Air valve opens only as far as needed to supply air demanded by engine.

Primary throttle partially open

LOW SPEED/LIGHT LOAD

Secondary throttle closed

HIGH SPEED/HEAVY LOAD

Secondary throttle opened by primary throttle linkage

SECONDARY THROTTLE OPERATION

Secondaries are controlled by these means:
- Mechanical.
- Diaphragm.
- Mechanical with velocity valve.
- Mechanical with air valve.

RPD carburetors use the latter two methods in the 4GC and Q-jet, respectively. A diaphragm opened the end carburetors on 3 X 2 setups for Olds, Pontiac and Chevrolet. And, the mechanical linkage alone was used on some 3 X 2 setups and the 4 X 1 setup on the 140-HP Corvair.

OPERATION

Mechanical—The secondaries are progressively opened by a direct link from the primaries: They don't begin opening until the primaries are open about 40°. Three factors ensure positive closure of the secondaries: a *return spring, a return link* from the primary throttle, and, because the throttles are offset on their shaft, *airflow* helps close them.

Diaphragm—Secondary throttles also can be opened by a vacuum-controlled diaphragm. The vacuum source originates from one of the primary venturis. At maximum engine vacuum, the secondaries are completely open, but they stay closed when the primaries are opened wide at low rpm.

Mechanical With Velocity Valve—This method is used in the 4GC. *Velocity valves* (blades) are placed above the secondary-throttle blades and below the rudimentary venturi structure. The secondary throttles are mechanically linked to the primaries; when the primaries open about 45°, the secondaries begin opening.

Little airflow passes the secondaries until the vacuum under the offset velocity valves is strong enough. The velocity valves delay the secondary airflow for smooth transition during secondary throttle valve opening. If engine rpm is high enough, the incoming air's velocity forces the spring-loaded velocity valves open. It takes one-half second to fully open the valves when the primary throttles are completely opened, *if the engine rpm is high enough.*

When full throttle is applied at low engine speeds, the velocity valves begin opening but close quickly because not enough air is flowing to force them open. Their spring tension limits opening until engine vacuum increases.

In the 4GC, the secondary side was sometimes equipped with an off-idle system. It offset a lean condition when the secondaries were cracked open at high-speed cruising, and helped eliminate slight bogging at their opening.

Mechanical with Air Valve—The Q-jet uses this method to control the secondaries. An *air valve*, which consists of a large plate, is mounted above the venturis. The secondaries are mechanically linked to the primaries; when the primaries open about 35°, the secondaries begin opening. No air flows through the secondaries until the vacuum under the offset air valve is strong enough to pull it open.

The air valve opens until it and the primaries can handle the airflow required by the engine. Its opening rate is controlled by a damping diaphragm and connecting rod so bogging does not occur.

Fuel is delivered for transitions—when the air valve starts to open—from ports just above or just below the valve. This action is similar to that of a secondary accelerator pump because it supplies a shot of fuel until the secondary main system takes over.

The secondaries also have a variable metering system that uses tapered metering rods in fixed metering orifices (jets). A cam on the air valve raises a metering-rod hanger attached to the rods. Opening the air valve lifts the tapered metering rods to reduce the jet diameter, allowing more fuel to flow. Refer to the Q-jet Secondary Metering Rods specification listing, page 152, to determine the metering-rod positions for metering orifices. This chart specifies metering area around the metering rod with various secondary air-valve openings.

There is no idle system in the secondary side of the Q-jet.

OTHER CARBURETOR FUNCTIONS

The carburetor also influences other engine subsystems, particularly those involving spark advance/retard and emission control. Signals created by various pressure areas in the carburetor, as related to throttle position, are applied as controls.

Vacuum-Advance Ports—Most carburetors have a slot or holes drilled above and alongside—not connected to!—the off-idle discharge slot or holes. These feed

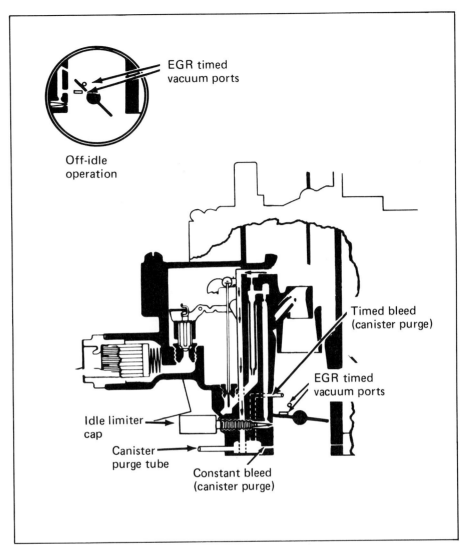

Some emissions control signals originate from carburetor.

vacuum to operate the distributor's vacuum advance when the throttle is opened. This vacuum-advance control is sometimes referred to as *timed spark*.

Canister Purging—Most 1970-and-later GM vehicles have closed fuel-tank venting to control evaporative emissions. The vent from the fuel tank leads to a vapor-collection canister. From 1972, all Q-jets have a bowl vent to this canister.

Because the fuel tank is not vented to the atmosphere and the carburetor is vented only to the canister when the engine is stopped, fuel vapor is collected in the vapor canister. Purge ports for the canister are in the carburetor. This plumbing varies by vehicle; more detail is given on page 168 in the Emissions Control chapter.

Exhaust Gas Recirculation (EGR)—From 1972, all GM models use an EGR system to reduce oxides of nitrogen emissions. The EGR valve is operated by a vacuum-supply signal taken from the carburetor. Its function is discussed in detail on page 166 in the Emissions Control chapter.

This chapter concentrates on explaining carburetion theory. The objective is to understand an engine's air/fuel (A/F) requirements so you can then apply a carburetor that satisfies them.

Don't worry about following complicated mathematical formulas here because I've distilled carburetion theory into some fundamental concepts. The complex calculations have been left to the engineers who design Rochesters.

METERING

An engine's performance depends on the air and gasoline mixture supplied to it. The weight ratio of air consumed to fuel burned in a given amount of time is termed the *A/F ratio*. For a gasoline engine, an ideal A/F ratio results in a maximum amount of energy. If the ratio varies from this value, less energy will be produced per pound of mixture. This ideal A/F ratio is approximately 15 pounds of air to 1 pound of fuel.

AIR-FLOW REQUIREMENTS

The air an engine consumes enters through the carburetor, so knowing how much air an engine can effectively use will help us select the correct carburetor.

FLOW RATINGS

Carburetor airflow rating in *cubic feet per minute*—cfm—is a more accurate measure of carburetor flow capacity than the older method of comparing venturi sizes. Venturi size doesn't represent the actual flow capacity of the carburetor, particularly because one or more booster venturis reduce the effective opening and increase the restriction or pressure drop across the carburetor.

NOTE: Displacements in cubic centimeters (cc) are divided by 16.4 to convert them to cubic inch displacement (CID) for the formulas.

Air/fuel ratios for each model must be met during computer flow bench inspection or carburetor is rejected for repair.

How big should a carburetor be? Two variables plugged into a simple formula can help determine the best-size carburetor for an engine. They are:

- *Engine Displacement*—specified in cubic inches.
- *Maximum rpm*—the peak rpm the engine will be run.

You must be realistic with these values. Inflated figures will cause you to buy too large a carburetor, which is a waste of time and money.

Now, plug these numbers into the following formula:

$$\frac{CID \times rpm \times \text{Volumetric Efficiency}}{3456} = cfm$$

Example: 350 CID
       7000 rpm maximum
       Assume volumetric efficiency
       of 1 (100%)

$$\frac{350 \times 7000 \times 1}{3456} = 709 \ cfm$$

A volumetric efficiency of 100% or 1 is usually not achieved with a naturally aspirated (unsupercharged) engine. So, unless supercharged, the example engine won't consume 709 cfm of air.

Volumetric Efficiency (VE)—This is a measure of how well the engine breathes. The better the breathing ability—the higher the volumetric efficiency. Volumetric efficiency is an incorrect description of *mass efficiency,* which is the value actually being measured. But its usage is established, so the term is used here.

Volumetric efficiency is defined as the ratio of the *actual* mass (weight) of air taken into the engine, to the mass the engine displacement would *theoretically* consume if there were no losses. This ratio is expressed as a percentage. VE is low at idle and at low rpm because the engine is throttled by the throttle-blade position.

VE reaches a maximum at an engine speed close to where maximum torque at WOT occurs, then falls off as engine speed is increased to peak rpm. The VE curve closely follows the engine's torque curve.

Different engine types produce various VE percentages. Ordinary low-performance engines have a VE of about 75% at maximum speed; about 80% at maximum torque. High-performance engines have a VE of about 80% at maximum speed; about 85% at maximum torque. A full-race engine has a VE of about 90% at maximum speed; about 95% at maximum torque.

A highly-tuned intake and exhaust system, matched with efficient cylinder-head porting and camshaft, can fill cylinders so completely that a VE of 100%—or slightly higher—may be obtained at the speed matching the system's tuned point or range.

Return to the example of the 350-CID engine we calculated flowing 709 cfm of air, at standard temperature and pressure, with 100% VE. If this were a high-performance engine with a maximum 80% VE, then the flow becomes 709 cfm X 0.80 = 567 cfm, at standard temperature and pressure.

AIR MASS & DENSITY

Because the mass of air ingested is directly related to its density, VE (η) can be expressed as a ratio of the density achieved in the cylinder to the inlet density, or:

$$\eta = \gamma \ cyl \div \gamma \ inlet$$

Ideal mass flow for a four-cycle engine is calculated by multiplying

$$\frac{rpm}{2} \times CID \times \gamma \ inlet$$

Air density varies directly with pressure—the lower the pressure, the less dense the air. Similarly, at altitudes above sea level, pressure drops, reducing power because density is reduced. Tables which related air density to pressure—correct barometer—and temperature are available. Or you can use this formula:

$$\gamma = \frac{1.326P}{t + 459.6}$$

where:
    γ = density in pounds per cubic foot.
    P = absolute pressure in in.Hg read from a barometer—not corrected.
    t = temperature in degrees F at the induction-system inlet.

Actual mass flow into an engine can be measured as the engine is running. A laminar-flow unit or other gas-measuring devices such as a calibrated orifice or a pitot tube do this.

Actual mass flow is usually lower than ideal in a naturally aspirated engine because air becomes less dense as it is heated in the intake manifold. Absolute pressure—and density—also drop as the mixture travels from the carburetor inlet to the combustion chamber. This further reduces the mass of the charge reaching the cylinder.

The greater the pressure drop through the carburetor, the lower the density inside the manifold and combustion chamber. If the carburetor is too small, the pressure drop at WOT will be too large. Power will be reduced because the mixture will not be as dense as necessary for full power.

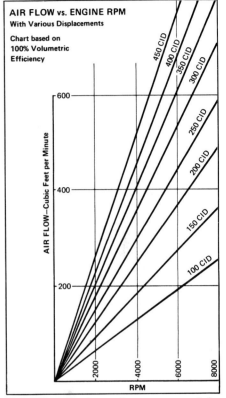

Pick CID, estimate maximum rpm, read airflow requirement, then correct airflow by VE as described in text.

Carburetors are flow-rated at WOT, and the cfm rating is the amount of airflow that causes a specified pressure drop, due to air friction through the carburetor. Rochester has rated their four-barrels at a drop of 1.5 in.Hg and other carbs at 3.0 in.Hg, as shown in the chart on page 108. The higher the flow rating, the bigger the carburetor. Or, the bigger the carburetor, the lower the pressure drop across it.

To keep VE as high as possible, keep the size of the carburetor up, thus the pressure drop down. However, the limiting factor is at the *low end* of the flow curve. Will the carburetor be able to meter fuel correctly at low airflows? Will it work well at the common speed range of the engine?

MORE AIR, MORE HP

If you're unfamiliar with thinking about the density of the combustion charge as it enters during the intake stroke, all this talk of mass flow, pressure drop, and volumetric efficiency probably becomes confusing. Here is an engine model to ponder which should make these concepts clearer.

The engine is an air pump—it ingests air and compresses it. It's often assumed the reason air enters the cylinders is because the piston sucks it in—it creates vacuum and that pulls in the air. But, atmospheric pressure—about 15 psi—will push air into a void, provided its passage is non-restrictive.

When a carburetor is large and its throttle is wide open, atmospheric pressure helps fill the low-pressure area above a piston on an intake stroke. Close the throttle to a small opening and atmospheric pressure forces a partial charge of air past the restriction. The result is less-dense air in the consuming cylinder, therefore less power.

Consequently, a small venturi and throttle-blade opening restrict the amount of air atmospheric pressure can push to the cylinder. For most street and highway driving, a small venturi carburetor is most efficient. If your goal is maximum power, however, and low- and mid-range economy, driveability and torque are secondary, then think big.

Think big all the way. Bigger intake ports, a higher-lift cam with longer duration, bigger intake valves and an efficient exhaust system to properly discharge burned mixture. Restrictions anywhere in the induction path keep atmospheric pressure from driving in a full air charge, so the engine will not perform to full potential.

The more air that atmospheric pressure can force in to the combustion chamber, the more HP.

FUEL REQUIREMENTS

Fuel requirements relate to the airflow requirement because fuel is consumed in proportion to the air being taken in by the engine. Fuel flow is stated in *pounds per hour* (lb per hr) and sometimes in pounds per HP/hr., which is also termed *specific fuel consumption* because it specifies how much fuel is used for *each* horsepower in one hour.

First, let's determine the WOT full-power fuel requirement for our example engine that needed 709 cfm of air at 7,000 rpm with 100% VE. Our final value will be the maximum amount of fuel this engine can consume with an airflow of 709 cfm.

To convert airflow into lb per hr:

cfm x 4.38 = airflow lb per hr.

Where 4.38 is a factor for 60F at one atmosphere—14.7 psi.

Multiply by the F/A ratio, which we will assume to be a typical full-power ratio of 0.077—A/F ratio of 13:1.

cfm x 4.38 x F/A = fuel flow lb per hr,

or

709 x 4.38 x 0.077 = 239 lb per hr @ 7000 rpm

This consumption will never be reached, however, because, like airflow, fuel flow must be reduced by VE—assume 0.80 in this example.

cfm x 4.38 x F/A x η = Fuel flow lb per hr.

or

709 x 4.38 x 0.077 x 0.80 = 191 lb per hr @ 7000 rpm

Fuel flow will be less at lower rpm. For instance, with this engine at 3,500 rpm, the flow will equal one-half—95 lb per hr—assuming WOT.

The accompanying chart shows the fuel flow for various engine sizes over typical rpm ranges. If you don't have this book handy for calculating fuel requirements before buying a fuel pump or fuel line, here is a rule of thumb.

WOT full power typically requires 0.5 lb fuel per HP per hour. Thus, a 300 HP engine needs 300 X 0.5 = 150 lb per hr. Gasoline weighs about 6 lb per gallon. So, 150 lb per hr divided by 6 lb per gallon equals 25 gallons per hr fuel flow at WOT.

Stoichiometric Mixture—This is the ideal fuel mixture—proportioned so all of the fuel burns with all of the air—the exhaust has only carbon monoxide, water and

---

nitrogen. This ideal F/A ratio is about 0.068, or an A/F ratio of 14.7:1.

Assuming an ideal set of conditions, the actual ratio at which this occurs varies with the fuel's molecular structure. Gasolines will vary somewhat in structure, but not significantly. Other fuels require different ratios for the stoichiometric condition.

Alcohol, for instance, has a lower heat content (calorific value) than gasoline and requires a 0.14 F/A ratio (7.15:1 A/F) for ideal burning. Because this fuel volume is so much more than that required for gasoline, most carburetors cannot be used with alcohol without being highly modified. A stock carburetor's passages are too small to correctly meter alcohol.

Maximum Power—This requires more fuel to combine with all the air. Excess fuel is required because mixture distribution to the cylinders and air/fuel mixing are seldom perfect. This excess usually amounts to 10—15%, giving a F/A ratio of 0.075—0.080, or a 13.3:1—12.5:1 A/F ratio. Sometimes, a F/A ratio larger than that which produces maximum power is used to internally cool an engine. Unburned hydrocarbons and carbon monoxide are undesirable products of this rich condition.

A good maximum power mixture is 12:1 because horsepower will be strong and a richer mixture aids combustion-chamber cooling to prevent detonation and pre-ignition. At about 7:1 the mixture no longer burns, and black smoke and extreme sluggishness occur.

Maximum Economy—Maximum economy requires excess air to the engine. If more air is admitted so the A/F ratio is greater than 15:1, air will be left after combustion is complete, but most of the gasoline will be burned. This lean ratio could

---

RELATING FUEL & AIR

The relationship between the amounts of fuel and air that flow into an engine is called the *fuel/air (F/A) ratio*. This value equals pounds of fuel divided by pounds of air. Engines use more air than fuel, so this ratio is always a small number—0.08, for example.

This relationship is also expressed as the *air/fuel (A/F) ratio*, which is another way to state the same value. It means the ratio is inverted and its value now equals pounds of air divided by pounds of fuel. So, a 0.08 F/A ratio equals a 12.5:1 A/F ratio. An accompanying table specifies these equivalent ratios.

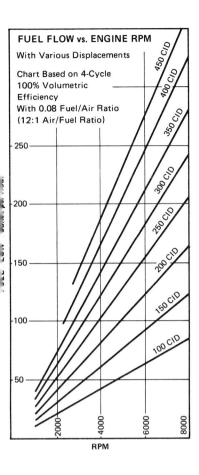

FUEL FLOW vs. ENGINE RPM

With Various Displacements

Chart Based on 4-Cycle
100% Volumetric
Efficiency
With 0.08 Fuel/Air Ratio
(12:1 Air/Fuel Ratio)

In this diluted mixture, some fuel molecules combine with exhaust molecules, and other fuel molecules align with the oxygen in the air. The mixture is made 10—20% richer than stoichiometric to offset the fuel combining with exhaust. The richer mixture helps offset distribution problems, but it creates emission problems because rich idle mixtures generate excessive carbon monoxide.

Carburetors built before 1980 supply a rich mixture in the range of 10:1—13:1 to overcome dilution. Microprocessor-controlled systems introduced in 1980 often supply an A/F ratio around 14:1—15:1. Beginning in the mid-'70s, carburetor manufacturers narrowed the mixture requirements of specific engines. Then only minor idle adjustments were permitted—idle-limiter devices were put on the idle-mixture adjustment screw.

Current mixture adjustment screws are sealed with a plug and it takes some effort to get access to them. For details, see page 97.

Cold Starting—Starting a cold engine requires the richest mixture of all because slow cranking speeds create air velocity that is too low to pull fuel droplets. Both the

fuel and manifold are cold, so vaporization is minimal. The fuel must be partially vaporized to get to the cylinders in a burnable condition.

Once the engine fires, speed goes up, velocity through the carburetor improves and vacuum increases. The result is better vaporization and the liquid fuel deposited on the manifold walls vaporizes. As the engine warms up and the choke comes off, the mixture is leaned out to normal.

EQUIVALENT RATIO TABLE

A/F (Air/Fuel)	F/A (Fuel/Air)	A/F (Air/Fuel)	F/A (Fuel/Air)
22:1	0.0455	13:1	0.0769
21:1	0.0476	12:1	0.0833
20:1	0.0500	11:1	0.0909
19:1	0.0526	10:1	0.1000
18:1	0.0556	9:1	0.1111
17:1	0.0588	8:1	0.1250
16:1	0.0625	7:1	0.1429
15:1	0.0667	6:1	0.1667
14:1	0.0714	5:1	0.2000

Power curve of typical engine relative to air/fuel mixtures. Curve is flat between air/fuel ratios 11:1 and 14:1, meaning engine will produce good horsepower in broad mixture range.

cause *engine surge*. In modern engines, if the A/F ratio approaches 20:1, the mixture is too lean to burn. Heavy surge, hesitation and backfire result. Some current engines, and earlier ones, begin objectionable lean operation in the 15:1—17:1 range.

Under heavy engine loads, with a mixture leaner than stoichiometric, there is sufficient heat to cause any free oxygen to combine with remaining nitrogen. These oxides of nitrogen (NOx) are some of the undesirable emission products. For more comment on this, refer to Emissions Control, page 158.

Idle—Little air and fuel enter the engine at idle. Because the pressure is extremely low in the cylinders, more exhaust gases remain in the combustion chamber. Therefore, a 15:1 mixture can leave the carburetor, but be diluted by residual exhaust gases in the combustion chamber until it can be as lean as 20:1. So, the engine runs poorly or stalls.

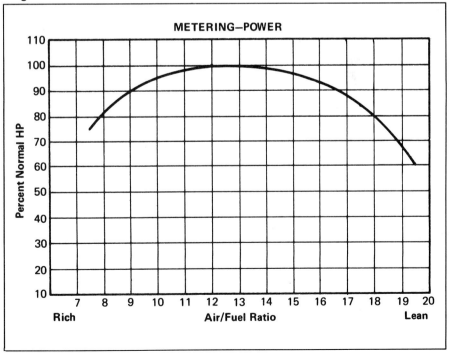

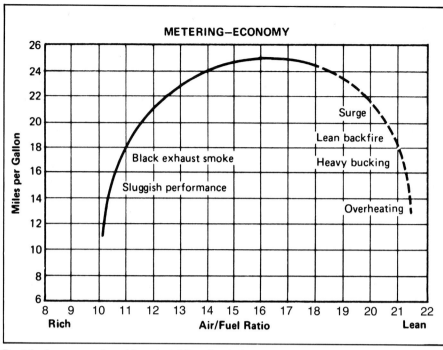

Best fuel economy is achieved at A/F ratio of about 16:1.

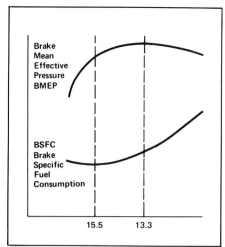

Brake mean effective pressure (BMEP) of engine operating with perfect distribution, showing relationship of A/F ratio supplied by carburetor. For maximum power, 13.3:1—15.5:1 A/F gives lowest possible fuel consumption. For maximum economy, 15.5:1 A/F gives lowest possible fuel consumption consistent with lower power.

UNIFORM DISTRIBUTION

Automobile-engine designers—and designers of carburetion systems—use different methods to check cylinder mixture. These methods require a chassis or engine dynamometer. Distribution is checked by:

• Chemically analyzing exhaust-gas samples from the individual cylinders at various operating ranges—idle, cruise, acceleration, maximum power and maximum economy. There is a definite relationship between A/F and the chemical components in the exhaust gas, as described in the Emissions Control chapter.

By using analytical equipment developed for emission studies, exact determinations of the A/F ratio are used to check distribution. Chemical analysis is the method preferred by automotive engineers for distribution studies.

• Measuring exhaust-gas temperature (EGT) with thermocouples inserted into the exhaust manifold or header at each cylinder's exhaust port. The designer looks for EGT peaks as various main-jet sizes are tried. If all cylinders were brought to the same temperature, one cylinder—or more—might be 200F (94C) below its

peak, and therefore below its peak output. The reason for this temperature difference is that all cylinders do not produce the same EGT due to differences in cooling, valves, porting and ring sealing. Perfect distribution would place all cylinders at their peak EGTs with the same jet—*an ideal that is rarely achieved.*

• Studying the brake specific fuel consumption to determine how closely the A/F requirements of a particular carburetor/manifold combination relate to ideal A/F ratios at various operating conditions. Any large variance is cause for suspecting a distribution problem.

• Observing combustion temperatures at various operating conditions.

FACTORS AFFECTING DISTRIBUTION

Atomization & Vaporization—Near-perfect distribution and best performance are attained if the mixture is atomized rather than separated fuel and air. Gasoline drops get momentum from the air stream in the manifold and are reluctant to turn corners. On the other hand, air responds to directional changes and turns corners easi-

ly. It meets cylinder demand—sometimes leaving gas drops on the manifold surfaces or at the corners. If the mixture isn't correctly atomized, cylinder efficiency will vary.

When vaporization is incomplete, liquid fuel gets in the cylinders. Because it does not completely burn, it is expelled as unburned hydrocarbons. Sometimes, the extra fuel washes oil off the cylinder walls, causing rapid wear. Liquid fuel can then drain past the worn rings into the crankcase and dilute the oil.

Intake Manifolds—Fuel distribution is affected by exhaust-heated hot spots in the manifold just under the carburetor. Hot spots are kept as small as possible to be consistent with the needs of flexible engine operation and smooth running. By keeping the spots small, the manifold automatically cools off as rpm increases. Fuel vaporized at high speed extracts heat from the manifold—often making it so cold that water condenses on its exterior.

Although most passenger-car intake manifolds heat the air/fuel mixture with an exhaust-heated spot, some manifolds are water-heated by the engine coolant. Cars

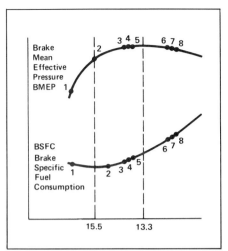

Same curve as at left with numbers identifying cylinders receiving A/F ratios at maximum-power output with non-perfect distribution. Cylinders 6,7,8 with too-rich A/F, produce less power and consume excess fuel; 3,4,5 receiving correct A/F for best power consume least fuel for that output; 1,2 have lean A/F ratio. Richer mixture would have to be supplied at ALL cylinders to bring 1 & 2 to flat part of curve and avoid detonation.

the Reid number varies from 8.5 in summer to 15 in winter. Fuel blending is a compromise based on estimates of what the temperature will be when the gasoline is used. A sudden temperature change—such as a warm day in winter—usually causes a rash of vapor lock and hot-starting problems.

Carburetor Placement—The carburetor's manifold location can drastically affect distribution. If the manifold layout places the carburetor closer to one or more cylinders, this can create distribution problems. Carburetor location becomes particularly critical when using multiple carburetors.

Combining Components—The engine designer checks power, economy and mixture distribution with the carburetor, air cleaner and manifold installed. Sometimes the positioning of the air cleaner or a connecting elbow atop the carburetor is so critical, that turning it to a different angle causes distribution problems.

Throttle-Plate Angle—Fuel direction is altered by a partially open throttle. This effect is apparent when the manifold has no riser between the carburetor and the manifold passages.

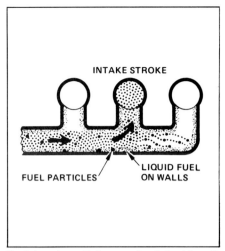

If mixture is partially vaporized, liquid particles clinging to manifold walls (or avoiding sharp turns into cylinder) may cause some cylinders to run lean and some rich. In this example, center cylinder will tend to receive a rich mixture.

equipped with emission controls often heat the incoming air by passing it over a "stove" on the exhaust manifold on its way to the air cleaner.

Passenger-car intake manifolds are compromises. Their shape, cross-sectional area and heating arrangements promote mixture distribution and volumetric efficiency over a range of engine speeds. If only maximum or near-maximum rpm is used, high mixture velocity through the manifold will help distribute and vaporize fuel—or at least hold the smaller particles of fuel in suspension in the mixture. At lower engine speeds, manifold heat is essential to vaporize fuel. If heat isn't used, the engine will run rough and distribution problems increase.

Fuel Composition—The more volatile the fuel, the better it vaporizes. Fuels are blends of hydrocarbon compounds and additives. Blending is done to match the fuel to ambient temperature and altitude conditions. Thus, fuels available in the summertime have a higher boiling point than those available in winter.

Fuel volatility is rated by *Reid Vapor Pressure numbers*. In Detroit, for example,

Webster-Hiese valve being developed by Doug Roe Engineering, Inc. helps distribution by improving fuel atomization.

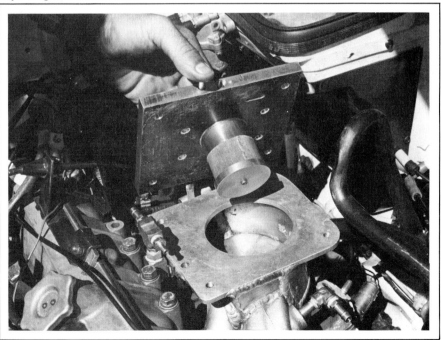

Throttle-plate angle affects distribution in the upper level of a cross-H (two-level) manifold—for a V8 engine—more than it does the lower level. The longer riser into the lower level straightens the mixture. This reveals one of the advantages of high-riser designs—they straighten mixture flow with less directional effect at all throttle openings.

High-riser designs better distribute the mixture to each cylinder. While height is usually limited by hood clearance, tests have shown improvements in distribution at part-throttle with 1-1/2—2-in.-high risers above manifold branches.

Mixture Speed & Turbulence—Mixture velocities and turbulence in the manifold affect vaporization and distribution. Proper turbulence in the combustion chamber helps to prevent stratification of the fuel and promotes rapid flame travel.

Time—Fuel volatility and heat available for vaporization are particularly important given the limited time available for vaporization. Unlike water in a tea kettle which can be left on a stove to boil—fuel must be vaporized in less than 0.008 second (8 milliseconds)—in a 12-in. manifold passage at a mixture velocity of 125 feet per second. This value is for an engine speed of 5000 rpm and a cam with 270 degrees duration. Whatever the parameters—that's fast!

DISTRIBUTION SUMMARY

- Vaporize as much fuel as possible in the manifold so a minimum of liquid fuel goes to the cylinders.
- Use fuels with the correct volatility for the ambient temperature and altitude.
- Keep velocities high in the manifold by using the smallest passages consistent with desired volumetric efficiency.
- Get superior atomization in the carburetor by accurately selecting venturi size (flow capacity).
- Avoid manifold construction that causes fuel to separate from the mixture due to sharp turns and severe changes in the cross-sectional area.
- Establish sufficient turbulence in the manifold to ensure fuel and air are well-mixed in the cylinder.
- To ensure fuel is well-vaporized, use manifold heating to heat the incoming air.

Cross section though V8 engine manifold of Cross-H or two-level type shows how riser height can affect distribution. Throttle shown in worst position causing greatest effect on distribution.

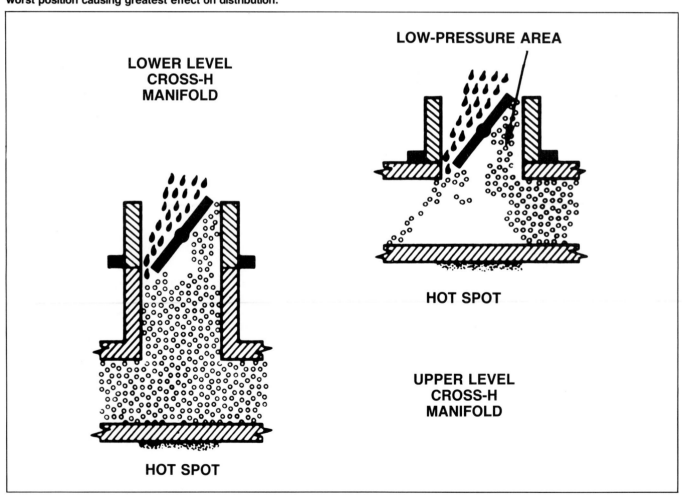

LOW-PRESSURE AREA

LOWER LEVEL
CROSS-H
MANIFOLD

HOT SPOT

UPPER LEVEL
CROSS-H
MANIFOLD

HOT SPOT

1974 Monojet for Vega used idle-stop solenoid as anti-dieseling device. Many other Monojets also used this feature.

EARLY DESIGNS

RPD has been a supplier of single-barrel carburetors since 1950. Metering sophistication became important during the '60s, RPD met this challenge with the Monojet. From the mid-'60s until 1979, it was the mainstay General Motors single-barrel on passenger cars and many light-duty trucks. It remained on some trucks through 1986. The metering controls in this carburetor were an advantage in meeting emission calibrations.

MONOJET

Monojets are designated by the letter M. Manual-choke models use only the M des-

ignation. Automatic-choke models also include the letter V to indicate the vacuum-break diaphragm on the carburetor and thermostatic coil on the engine. So, automatic-choke models are labeled MV.

The Monojet is a single-bore downdraft carburetor with a tube nozzle used with a multiple venturi. Fuel flow through the main-metering system is controlled by a mechanically and vacuum-operated variable-area jet. A tapered rod operating in the fixed-orifice main-metering jet is directly connected by linkage to the main throttle shaft. The same rod is also vacuum-operated by a power piston to form a

power-enrichment system with the main-metering system. This dual arrangement improves performance during moderate-to-heavy acceleration.

An exception to this metering design is found on the Monojet used on the lowest-HP Vega engine. In place of the metering rod and vacuum-operated power system, it has a conventional metering jet with a velocity-actuated power-enrichment valve. This design was influenced by economy goals for the small-displacement engine. It's a prudent choice in my opinion, because the acceleration and cruise vacuum are nil, which would make a vacuum-

29

Model MV Monojet and emission-control equipment with external vacuum break

1   Air horn assembly
2   Air horn—long screw
3   Air horn—short screw
4   Air cleaner stud bracket
5   Bracket attaching screw
6   Air horn gasket
7   Choke shaft & lever assembly
8   Choke valve
9   Choke valve screw
10 Choke vacuum break unit
11 Vacuum break hose
12 Vacuum break lever
13 Vacuum break link
14 Vacuum break lever screw
15 Choke lever
16 Choke rod
17 Fast idle cam
18 Cam attaching screw
19 Float bowl assembly
20 Idle tube assembly
21 Main metering jet
22 Pump discharge ball
23 Pump discharge spring
24 Pump discharge guide
25 Needle and seat assembly
26 Needle seat gasket
27 Idle compensator assembly
28 Idle compensator gasket
29 Idle compensator cover
30 Cover screw
31 Float assembly
32 Float hinge pin
33 Power piston assembly
34 Power piston spring
35 Power piston rod
36 Metering rod & spring assembly
37 Fuel inlet filter nut
38 Filter nut gasket
39 Fuel inlet filter
40 Fuel inlet spring
41 Idle stop solenoid
42 Pump assembly
43 Pump actuating lever
44 Pump return spring
45 CEC valve
46 CEC vacuum tube
47 CEC valve nut
48 CEC valve bracket
49 CEC valve bracket screw
50 Throttle body assembly
51 Throttle body gasket
52 Idle needle limiter cap
53 Idle needle
54 Idle needle spring
55 Throttle body screw
56 Pump and power rods lever
57 Lever attaching screw
58 Power piston rod link
59 Pump lever link

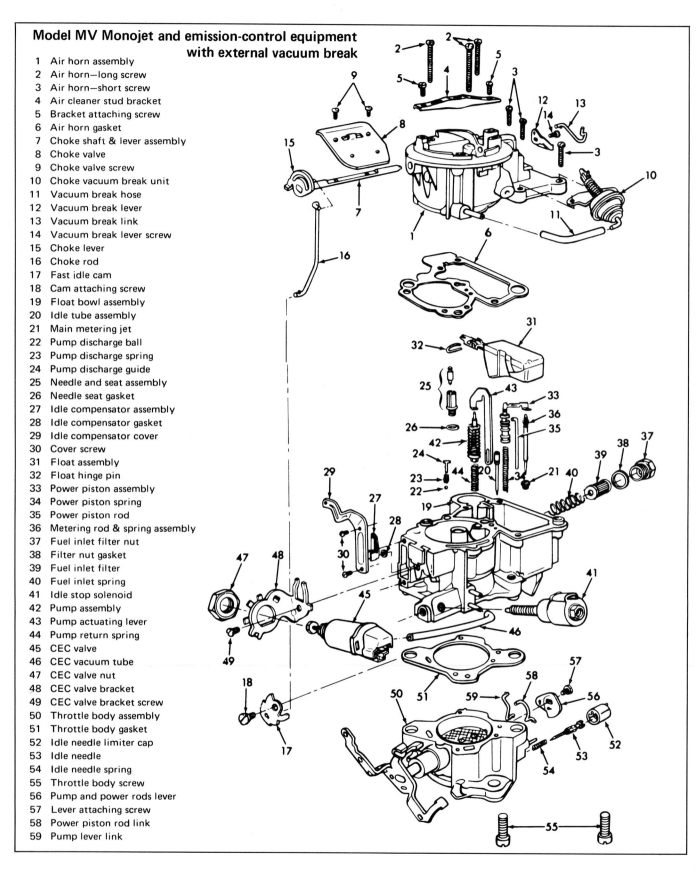

controlled system call prematurely for fuel enrichment.

Other Monojet features are:

• Aluminum throttle body for decreased weight.

• Thick throttle-body-to-bowl insulator gasket to protect the float bowl from engine heat.

• Internally-balanced venting hole in the air horn; an external idle-vent valve was also used on some pre-'71 models.

The simple design of the Monojet carburetor lends itself to easy servicing. A few illustrations here show how its system features differ from other early Rochester carburetor designs.

Main Metering System—The main system is augmented by an adjustable-flow feature. It is set to control part-throttle fuel mixtures more accurately than a fixed orifice. Adjustable-part-throttle (APT) fuel is channeled from the fuel bowl to the nozzle feed well, independent of the metering jet, but parallel with it.

Power System—The power system is integral to the main metering system—both operate on the same tapered metering rod in the main metering jet. At part-throttle and cruising, high manifold vacuum holds the power piston against spring tension. The upper side of the power-piston groove is held against the top of the drive rod so the metering rod is kept low in the jet for maximum economy.

Acceleration reduces manifold vacuum, then the power-piston spring pushes the piston up until the lower edge of its groove is against the bottom of the drive rod. This action raises the metering rod so more fuel flows through the main jet.

Power enrichment is controlled in most Monojets by clearance between the power-piston groove and the outside diameter (OD) of the drive-rod.

MODEL B

The B-series single-barrel was fitted to most Chevrolet six-cylinder engines produced during 1950—67. Letter designations denote choke variations: B has a manual choke and was used only on trucks; BC has a thermostat choke mounted on the air horn and was used on cars and trucks, and BV operates with a remote thermostatic coil on the exhaust manifold.

Its design features include the circular float bowl and the metering passages in the air horn. The float bowl encircles the ven-

B models were main A/F mixers for 1950—67 six-cylinder GM cars. Circular fuel-bowl design keeps fuel in throttle bowl during abrupt maneuvers.

"Snatch-idle" system pulls fuel across top of throttle bore. Idle fuel is "snatched" across main-discharge nozzle.

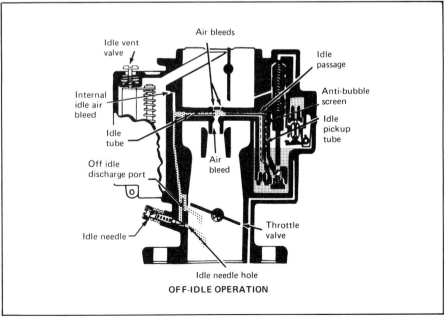

OFF-IDLE OPERATION

Models B, BC, BV—Typical Assembly

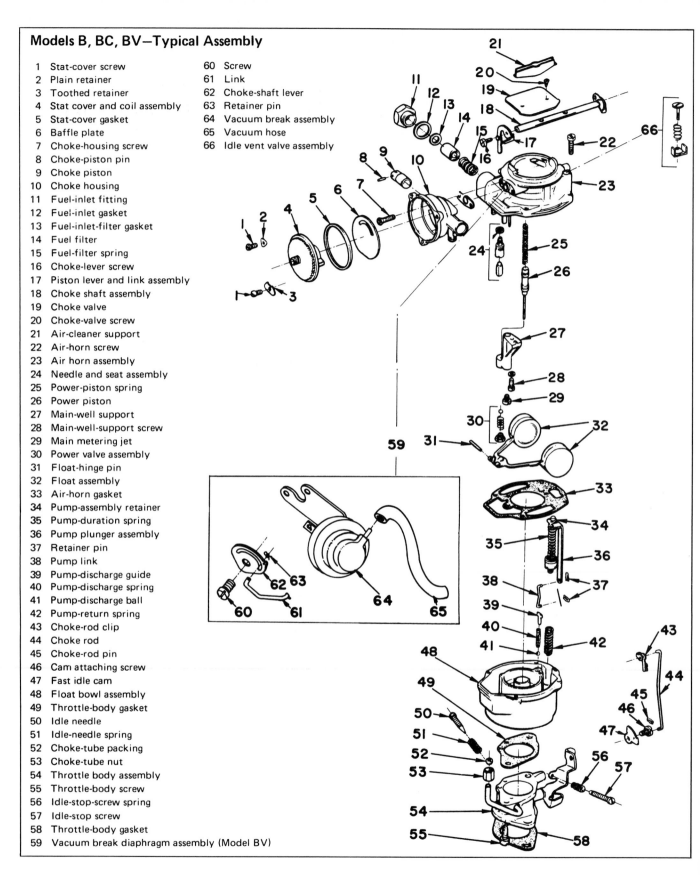

1 Stat-cover screw
2 Plain retainer
3 Toothed retainer
4 Stat cover and coil assembly
5 Stat-cover gasket
6 Baffle plate
7 Choke-housing screw
8 Choke-piston pin
9 Choke piston
10 Choke housing
11 Fuel-inlet fitting
12 Fuel-inlet gasket
13 Fuel-inlet-filter gasket
14 Fuel filter
15 Fuel-filter spring
16 Choke-lever screw
17 Piston lever and link assembly
18 Choke shaft assembly
19 Choke valve
20 Choke-valve screw
21 Air-cleaner support
22 Air-horn screw
23 Air horn assembly
24 Needle and seat assembly
25 Power-piston spring
26 Power piston
27 Main-well support
28 Main-well-support screw
29 Main metering jet
30 Power valve assembly
31 Float-hinge pin
32 Float assembly
33 Air-horn gasket
34 Pump-assembly retainer
35 Pump-duration spring
36 Pump plunger assembly
37 Retainer pin
38 Pump link
39 Pump-discharge guide
40 Pump-discharge spring
41 Pump-discharge ball
42 Pump-return spring
43 Choke-rod clip
44 Choke rod
45 Choke-rod pin
46 Cam attaching screw
47 Fast idle cam
48 Float bowl assembly
49 Throttle-body gasket
50 Idle needle
51 Idle-needle spring
52 Choke-tube packing
53 Choke-tube nut
54 Throttle body assembly
55 Throttle-body screw
56 Idle-stop-screw spring
57 Idle-stop screw
58 Throttle-body gasket
59 Vacuum break diaphragm assembly (Model BV)

60 Screw
61 Link
62 Choke-shaft lever
63 Retainer pin
64 Vacuum break assembly
65 Vacuum hose
66 Idle vent valve assembly

turi which, combined with the centrally located nozzle, prevents fuel spill-over during quick turns or stops. All metering passages—except for the idle passages—are in the air horn. Because these passages are kept cooler by the air-horn gasket and gas in the float bowl, the B models have steady metering consistency and less chance for vapor lock than previous RPD carburetors.

The B's most unusual feature is the *snatch-idle*. The idle tube is located across from the main fuel system. Idle fuel is "snatched" across the air gap into the idle-down passage. One side benefit of this design is that the main-nozzle passage is always primed and ready to supply fuel on nozzle demand. Consequently, a small-capacity accelerator pump with a short stroke can be used because there is less lag in main-system fuel start-up.

MODEL H

This carburetor has a unique history: it was designed, developed and produced exclusively for one vehicle—the Corvair. It used two of these carburetors connected by a throttle-rod and air-cleaner arrangement. Some models employ four H carburetors (4 X 1) with a progressive linkage.

The H carburetor was designed with many innovations and distinctions; only the foremost are described here.

The throttle body and float bowl are united into a single aluminum casting. This simplified maintenance in the field and reduced initial building costs. Repair kits were inexpensive and rebuild time was less than that required for other carburetors.

The fuel discharge assembly was a new design. A radial discharge nozzle replaced the conventional boost venturi. It has four "spokes" or arms to discharge fuel near the venturi surface. It is unusual, but it works.

Main-well-tube inserts are used with the main-well tubes for improved hot-engine idle stability and to prevent fuel percolation and general hot-starting problems.

Early H carburetors do not have a conventional power system. Pre-'65 Corvairs rely on pulse enrichment supplied by sharply defined pulses created by carburetor placement, manifold design and firing order. As exhaust-emission regulations increased the need for leaner light-throttle metering, a power system was added to 1965 models. It enriches mixture only during high inlet-air velocities. A similar

Model Hs are used in dual (2 x 1) or four-carburetor (4 x 1) installations on Corvairs.

Radial discharge nozzle replaced conventional venturi in Model H.

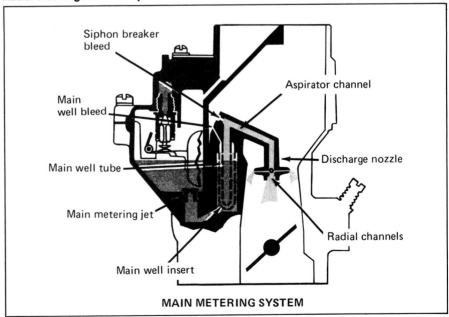

Siphon breaker bleed

Main well bleed

Aspirator channel

Main well tube

Discharge nozzle

Main metering jet

Radial channels

Main well insert

MAIN METERING SYSTEM

Models H, HV—Typical Exploded View

1   Choke-valve screw
2   Choke valve
3   Choke shaft and lever assembly
4   Fuel-inlet nut
5   Fuel-inlet-filter gasket
6   Fuel-inlet-nut gasket
7   Fuel-inlet filter
8   Fuel-inlet-filter spring
9   Air horn assembly
10   Air-horn screw (short)
11   Lockwasher-air horn screws
12   Choke lever and collar assembly
13   Trip lever
14   Trip-lever screw
15   Pump shaft and lever assembly
16   Needle-seat gasket
17   Float needle seat
18   Pump-lever inside springclip
19   Pump inside lever
20   Vacuum break control assembly
21   Air-horn screw (long)
22   Control-rod clip
23   Vacuum control rod
24   Upper pump-rod retainer
25   Float-hinge pin
26   Torsion spring
27   Float needle
28   Float assembly
29   Vacuum-control hose
30   Pump-plunger clip
31   Pump assembly
32   Pump-return spring
33   Air-horn gasket
34   Venturi-cluster screw (short)
35   Venturi-cluster screw (long)
36   Cluster-screw lockwasher
37   Venturi-cluster assembly
38   Venturi-cluster gasket
39   Main-well-tube insert
40   Power valve
41   Pump-discharge valve
42   Main metering jet
43   Body and bowl assembly
44   Pump rod
45   Lower pump-rod retainer
46   Pump-lever attaching screw
47   Pump actuating lever
48   Choke rod
49   Fast idle cam
50   Cam attaching screw
51   Slow idle screw spring
52   Slow idle adjustment screw
53   Idle needle spring
54   Idle needle
55   Idle-vent valve
56   Idle-vent attaching screw

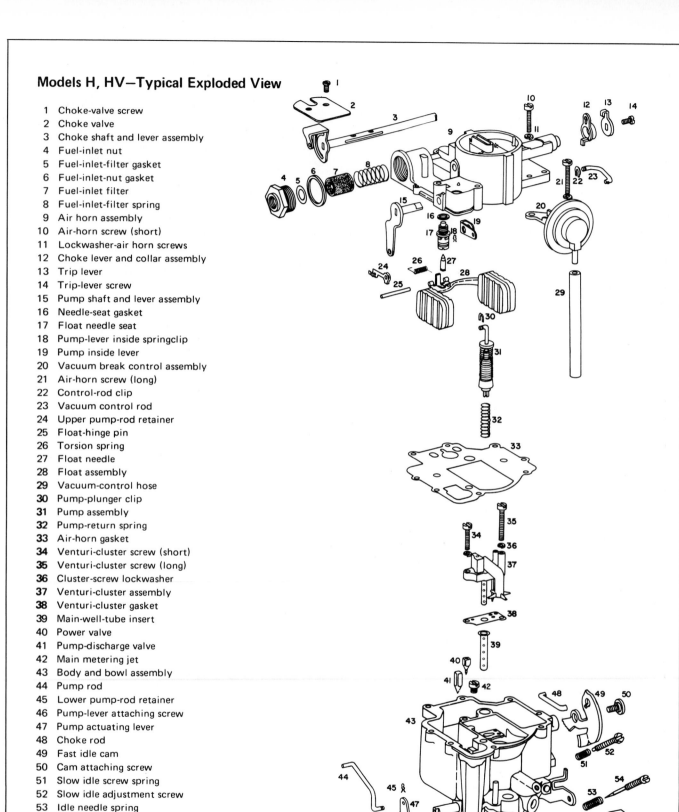

system is used in the Monojet for the Chevrolet Vega engine.

THE 2G

The 2G's versatility is second only to the Q-jet. Three Rochester two-bore G models were built: 2G, 2GC and 2GV. The basic model designation is G. The 2 indicates two bores, two venturis and two separate identical metering systems: one per bore. The 2G was used on V8 engines; each bore supplied mixture to four cylinders through a divided two-stage intake manifold.

The basic 2G is equipped with a manual choke. It was fitted on truck and marine engines that didn't require an automatic choke.

Model 2GC is the 2G with an automatic choke—designated by C—integral to the carburetor. The automatic-choke housing and thermostatic coil may be located on the air horn or throttle body, depending on the application.

Model 2GV also has an automatic choke. A vacuum-break diaphragm unit replaces the conventional choke housing and choke piston. The choke thermostatic coil is located on the exhaust manifold and line-connected to the choke valve.

This carburetor was designed for easy servicing. Most of the calibrated metering parts are in the venturi-cluster assembly. This simplicity helped establish an outstanding reliability record for RPD's two-barrel carburetors from the original design introduction in 1955. The last production units for passenger cars were built in 1978. Emission and fuel economy requirements forced their replacement by more elaborate designs.

THE 4G

Some of these are still seen today at older car drag races and on restorations. The 4G/4GC was RPD's first four-barrel design; it was used on GM cars from 1952—67. Two mid-'50s trends provoked this design—lower hood lines and demand for more power. So, it became necessary to shorten the carburetor and increase its capacity—the four-barrel era began.

The four-barrel carburetor was essentially two, 2-barrel carburetors in a single casting. The primary or fuel-inlet side contains all the systems for carburetion. The secondary side supplements the primary with separate float and power systems to add extra capacity for high-speed

Vacuum-break diaphragm design on 2GV.

1974 Chevrolet 2GV features teflon-coated throttle shaft for low friction. Electrical idle-stop solenoid maintains curb-idle speed.

MODELS 2G, 2GC, 2GV—Exploded View

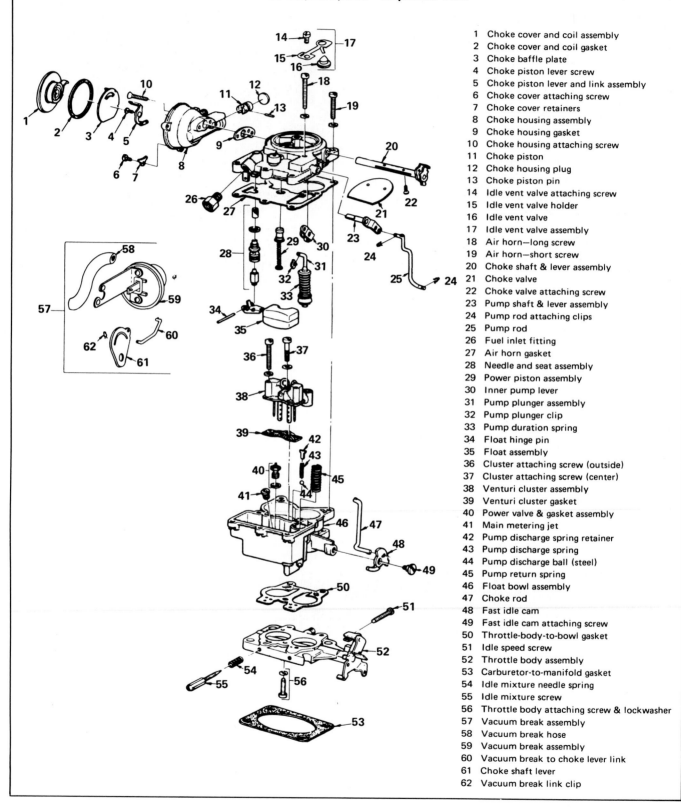

1. Choke cover and coil assembly
2. Choke cover and coil gasket
3. Choke baffle plate
4. Choke piston lever screw
5. Choke piston lever and link assembly
6. Choke cover attaching screw
7. Choke cover retainers
8. Choke housing assembly
9. Choke housing gasket
10. Choke housing attaching screw
11. Choke piston
12. Choke housing plug
13. Choke piston pin
14. Idle vent valve attaching screw
15. Idle vent valve holder
16. Idle vent valve
17. Idle vent valve assembly
18. Air horn—long screw
19. Air horn—short screw
20. Choke shaft & lever assembly
21. Choke valve
22. Choke valve attaching screw
23. Pump shaft & lever assembly
24. Pump rod attaching clips
25. Pump rod
26. Fuel inlet fitting
27. Air horn gasket
28. Needle and seat assembly
29. Power piston assembly
30. Inner pump lever
31. Pump plunger assembly
32. Pump plunger clip
33. Pump duration spring
34. Float hinge pin
35. Float assembly
36. Cluster attaching screw (outside)
37. Cluster attaching screw (center)
38. Venturi cluster assembly
39. Venturi cluster gasket
40. Power valve & gasket assembly
41. Main metering jet
42. Pump discharge spring retainer
43. Pump discharge spring
44. Pump discharge ball (steel)
45. Pump return spring
46. Float bowl assembly
47. Choke rod
48. Fast idle cam
49. Fast idle cam attaching screw
50. Throttle-body-to-bowl gasket
51. Idle speed screw
52. Throttle body assembly
53. Carburetor-to-manifold gasket
54. Idle mixture needle spring
55. Idle mixture screw
56. Throttle body attaching screw & lockwasher
57. Vacuum break assembly
58. Vacuum break hose
59. Vacuum break assembly
60. Vacuum break to choke lever link
61. Choke shaft lever
62. Vacuum break link clip

4Gs still get run at drags. Nova uses 4GC on 327 small block.

4GC is readily identified by forest of vents atop air horn. It was RPD's first four-barrel and used from 1952—1967. Note water hose connection (A) for choke thermostatic coil housing and adjustable accelerator pump lever (B) with five holes.

and power ranges.

It was considered efficient design to use as small a venturi as possible and keep high air velocity and have efficient metering. These design concepts were applied even more rigorously to the Q-jet.

By combining two carburetors into a single casting and delaying the opening of the second carburetor, the carburetor's capacity can be used as required. During normal driving, the primary side does the majority of the work and venturi sizes are kept small for efficient metering.

Between 42° and 60° primary-throttle opening, secondary throttles begin opening and their linkage is such that when the primaries reach full-open, the secondaries are also fully open. The secondary bores also contain a spring-loaded auxiliary throttle valve that further controls metering.

Special features of the 4G are shown in accompanying photos and illustrations.

4G ECONOMY TUNING
You have an obstacle to face when tuning your 4GC for utmost economy—no power-valve springs or assemblies are available because it is obsolete. You'll have to make any modifications to the original parts or to whatever is included in rebuild kits.

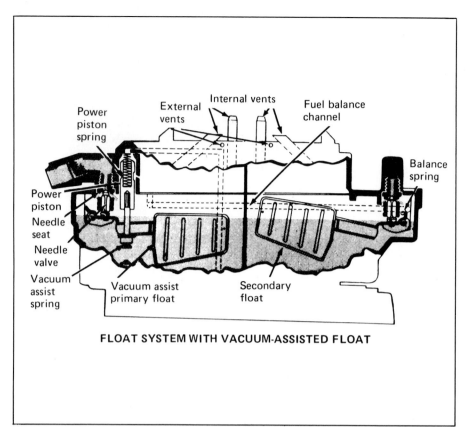

FLOAT SYSTEM WITH VACUUM-ASSISTED FLOAT

Power piston spring
External vents
Internal vents
Fuel balance channel
Balance spring
Power piston
Needle seat
Needle valve
Vacuum assist spring
Vacuum assist primary float
Secondary float

Air flow opens valve.
Calibrated spring (holds valve closed)
AUXILIARY THROTTLE VALVE

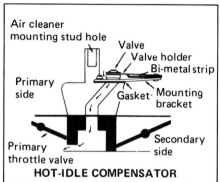

Air cleaner mounting stud hole
Valve
Valve holder
Bi-metal strip
Mounting bracket
Gasket
Primary side
Primary throttle valve
Secondary side
HOT-IDLE COMPENSATOR

4G features: Float system, hot-idle compensator and auxiliary throttle valve.

Spring tension of auxiliary throttle valves in secondary side can be adjusted with slim screwdriver. Special screwdriver (arrow) with 10° markings on circular handle is used here. Turn it counterclockwise to add tension after neutral position is established. Add tension in 10° increments until desired performance is obtained. Don't forget to tighten setscrew.

Models 4G, 4GC—Exploded View

1 Intermediate choke rod clip (lower)
2 Intermediate choke rod
3 Intermediate choke rod clip (upper)
4 Choke shaft & lever assembly
5 Choke valve screw
6 Choke valve
7 Air horn assembly
8 Air horn screw
9 Choke lever and collar assembly
10 Choke trip lever
11 Choke trip lever screw
12 Pump shaft & lever assembly
13 Pump rod clip
14 Pump rod
15 Pump shaft & lever clip
16 Air horn gasket
17 Fuel inlet fitting
18 Needle & seat assembly (primary)
19 Power piston assembly
20 Pump plunger boot
21 Pump plunger clip
22 Pump plunger assembly
23 Needle & seat assembly (secondary)
24 & 24A   Float assembly
25 Float balance spring & clip assembly
26 Float hinge pin
27 Venturi cluster screw & lockwasher
28 Venturi cluster (primary)
29 Venturi cluster (secondary)
30 Venturi cluster gaskets
31 Idle compensator assembly
32 Pump discharge guide
33 Pump discharge spring
34 Pump discharge check ball
35 Power valve assembly
36 Main metering jet (primary)
37 Main metering jet (secondary)
38 Pump inlet screen retainer
39 Pump inlet screen
40 Pump return spring
41 Pump inlet check ball
42 Float bowl assembly
43 Auxiliary throttle valve assembly
44 Throttle-body-to-bowl gasket
45 Idle speed screw
46 Idle speed screw spring
47 Throttle body assembly
48 Choke rod
49 Choke rod attaching clips
50 Fast idle cam
51 Fast idle cam screw
52 Fast idle screw spring
53 Fast idle screw
54 Throttle-body-to-bowl screw (large)
55 Throttle-body-to-bowl screw (small)
56 Idle mixture screw
57 Idle mixture screw spring
58 Choke cover & coil assembly
59 Choke cover & coil gasket
60 Choke baffle plate
61 Choke cover attaching screws
62 Choke cover retainers (toothed)
63 Choke piston lever screw
64 Choke piston lever and link assembly
65 Choke housing attaching screw
66 Choke cover retainer
67 Choke housing plug

68 Choke Piston
69 Choke piston pin
70 Choke housing assembly
71 Intermediate choke shaft & lever
72 Choke housing gasket
73 Carburetor-to-manifold gasket
74 Idle vent valve assembly
75 Filter element relief spring
76 Filter element
77 Filter element gasket
78 Fuel inlet fitting gasket
79 Fuel inlet fitting

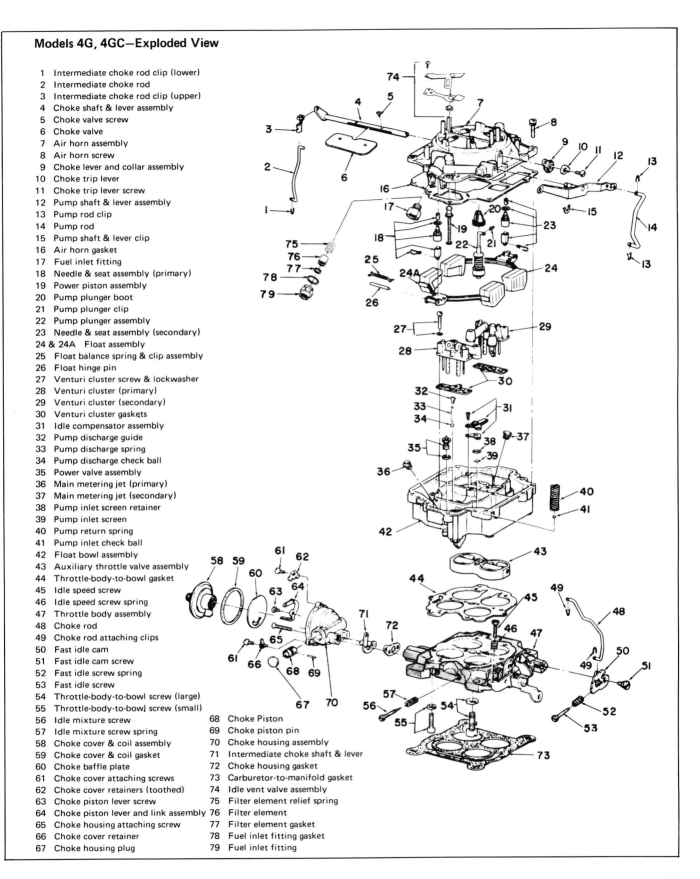

39

Early design 1978—80 Dualjet.

First-design non-electronic Varajet looks straightforward.

Later Varajet with electronic control. Not so simple now.

ROCHESTER DESIGNS FOR THE 1980s

Since the first printing of this book, new Rochester carburetors have been designed and put in production. In the Quadrajet Design chapter, I discuss the changes incorporated in the Q-jet so it could meet the emission-control and fuel-economy demands of the '80s. Rochester two-barrels have been replaced by more sophisticated units too.

DUALJET

One carburetor introduced in 1976 was a Q-jet *sans* secondaries. It was logical it would evolve when the quest for improved carburetion started in the mid-'70s. The primary side of a Q-jet was efficient and the two-barrel 2GCs of that era were not. So, to meet emission standards and strive for improved fuel economy, half a Q-Jet—the Dualjet—was put in service.

Model 200—In 1977, Pontiac introduced an M2MC Dualjet 200 on their 301 V8s.

This engine/carburetor combination was also used in 49-state Buick LeSabre vehicles. This unit uses a separated main well in the main metering system, triple venturi, and 1-3/8-in. bores.

Model 210—In 1978, some 1-7/32-in. throttle bores were introduced, making the Dualjet even more adaptable to smaller engines. This was identified as the Model 210. The 305 V8 is the largest engine that uses a Dualjet, but it's use was widespread throughout GM vehicles for a number of years in the '70s.

Dualjet Design—This is not an exciting carburetor. It supplies good, but not excellent, mixture ratios to engines during operation. It works best during the idle, off-idle—and through the main system—part-throttle ranges.

It is compromised when the main metering is jetted lean for optimum cruise and light-throttle operation. Then, the low-speed, medium-to-heavy acceleration suffers from leanness.

When electronic fuel metering was put on Dualjets in the early '80s, options for altering metering ceased. A microprocessor controls the carburetor's mixture ratio. If metering changes are made, it will readjust variables and establish the same mixture ratio in nearly all driving ranges.

Dualjet Service—All in all, the Dualjet gives excellent service with a minimum of repair. It is reliable if supplied reasonable service and not tinkered with. It is acceptably efficient as sold.

Read the Quadrajet Service chapter for general service procedures for the Dualjet. Much of what is detailed there applies; specific adjustments and settings for a particular vehicle will be in its service manual.

2SE VARAJET

Fuel economy standards enacted in 1978

1. Air Horn Assembly
2. Air Horn Gasket
3. Pump Actuating Lever
4. Pump Lever Hinge Pin
5. Air Horn Screw (Short)
6. Air Horn Countersunk Screw (2)
7. Solenoid-to-Air Horn Gasket.
8. Idle Air Bleed Valve
9. "O" Ring (Thick)
10. "O" Ring (Thin)
11. TPS Actuator Plunger
12. TPS Plunger Seal
13. TPS Seal Retainer
14. TPS Adjusting Screw
15. TPS Screw Plug
16. Pump Plunger Seal
17. Pump Seal Retainer
18. Solenoid Plunger Rich Stop Screw
19. Solenoid Plunger Rich Stop Screw Plug
20. Solenoid Lean Mixture Screw Plug
21. Front Vacuum Break and Bracket
22. Front Vacuum Break Screws
23. Vacuum Hose
24. Upper Choke Rod Lever
25. Choke Lever Screw
26. Choke Rod
27. Lower Choke Rod Lever
28. Intermediate Choke Shaft Seal
29. Rear Vacuum Break Link (If Equipped)
30. Intermediate Choke Shaft and Lever
31. Fast Idle Cam
32. Hot Air Choke Housing-to-Bowl Seal
33. Choke Housing Kit
34. Choke Housing-to-Bowl Screw
35. Intermediate Choke Shaft Seal (Hot Air Choke)
36. Choke Coil Lever
37. Choke Coil Lever Screw
38. Hot Air Choke Stat Cover Gasket
39. Hot Air Choke Stat Cover and Coil Assy.
40. Electric Choke Stat Cover and Coil Assy.
41. Stat Cover Retaining Kit
42. Rear Vacuum Break Assy. (If Equipped)
43. Rear Vacuum Break Screws
44. Float Bowl Assembly
45. Primary Metering Jet (2)
46. Pump Discharge Ball
47. Pump Discharge Ball Retainer
48. Pump Well Baffle
49. Needle and Seat Assembly
50. Float Assembly
51. Float Hinge Pin
52. Primary Metering Rod (2)
53. Primary Metering Rod Spring (2)
54. Float Bowl Insert
55. Bowl Cavity Insert
56. Connector Attaching Screw
57. Mixture Control Solenoid and Plunger Assembly
58. Solenoid Tension Spring
59. Solenoid Lean Mixture Adjusting Screw
60. Solenoid Adjusting Screw Spring
61. Pump Return Spring
62. Accelerator Pump Assy.
63. Pump Link
64. Throttle Position Sensor (TPS)
65. TPS Tension Spring
66. Fuel Inlet Filter Nut
67. Filter Nut Gasket
68. Fuel Inlet Filter
69. Fuel Filter Spring
70. Idle Stop Screw
71. Idle Stop Screw Spring
72. Idle Speed Solenoid (ISS) and Bracket Assy. (If Equipped)
73. Throttle Return Spring Bracket
74. Idle Load Compensator (ILC) and Bracket Assy. (If Equipped)
75. Idle Speed Control (ISC) and Bracket Assy. (If Equipped)
76. Bracket Screw
77. Throttle Body Assembly
78. Throttle Body Gasket
79. Throttle Body Screw
80. Idle Needle and Spring Assembly (2)
81. Fast Idle Adj. Screw
82. Fast Idle Screw Spring
83. Vacuum Hose "T"
84. Flange Gasket

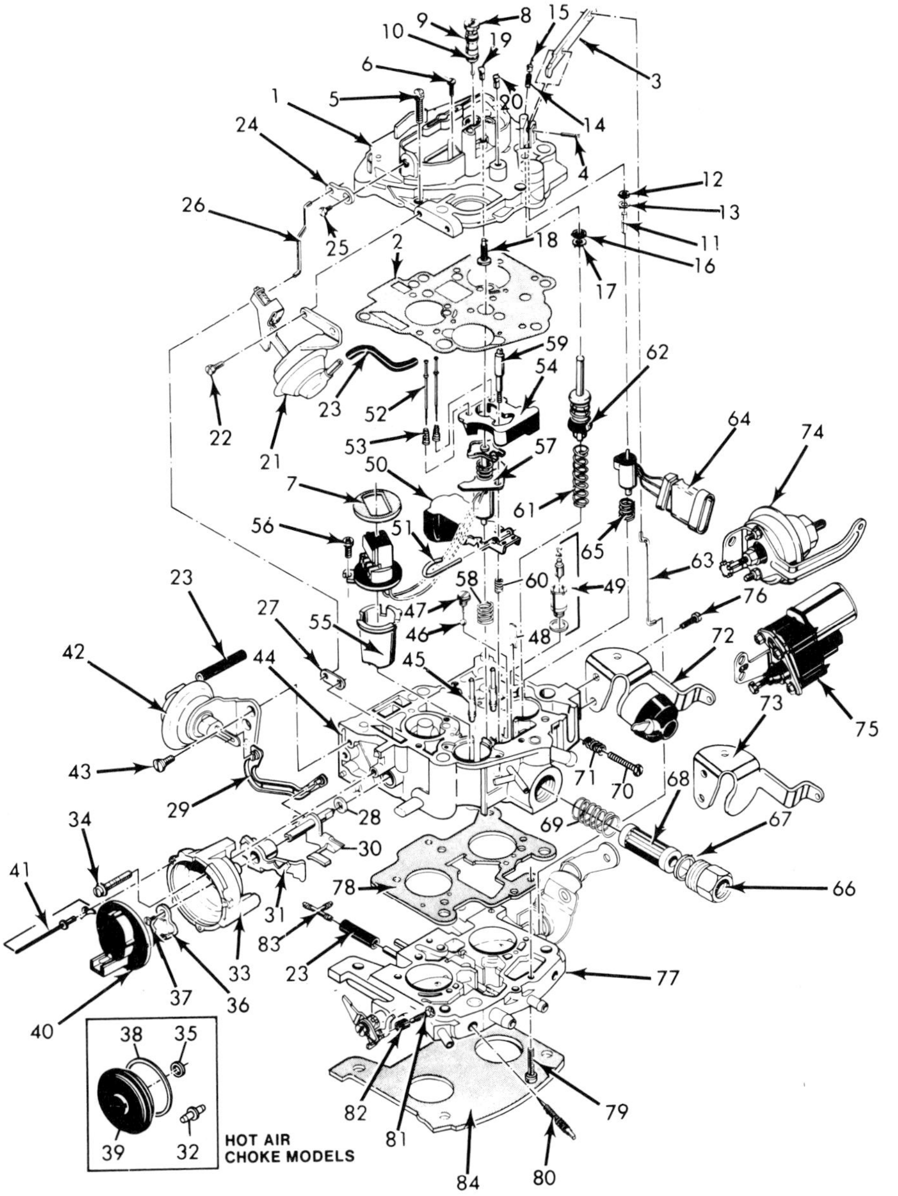

HOT AIR CHOKE MODELS

Dualjet (with electronic control) wasn't exciting design, but it got the job done. Drawing courtesy GM.

forced the realization that smaller engines in down-sized cars were a must. The Dual-jet's utility was being stretched to the limit on medium-size, low-performance GM vehicles.

A new carburetor was designed for smaller engines. This two-stage carburetor, built almost entirely of aluminum, first came out on GM front-wheel-drive vehicles. The Varajet would serve all GM divisions using the 2.5-liter four-cylinder and the 2.8 liter V6 until throttle body and port fuel injection technology made conventional carburetors obsolete.

Varajet Design—The first model (1979) 2SE didn't have a Computer Command Control (CCC) system with a mixture-control solenoid and electronic idle-speed control. It was introduced as a conventional downdraft carburetor. The electronics were added within two years.

Why did RPD go to the expense of designing a new model for small engines when they already had the Monojet with its good track record? For the same reasons the Dualjet replaced the 2G—the idle, off-idle and main system metering couldn't be controlled enough to meet emission and economy standards. If the Monojet were down-sized enough, and the venturi made sensitive enough to meet emission and fuel-economy standards, it would have been too small to deliver adequate HP.

The 2SE two-stage was designed with a 35-mm bore for fuel metering control during idle and part-throttle operation. That bore was too small to expect any reasonable HP, so a 46-mm secondary bore was added. This design supported the power requirements at heavy throttle.

An air valve is used in the secondaries with a single, tapered metering rod. Metering control is governed by the air-valve opening so a suitable power mixture prevails regardless of how far open the secondary is.

Another feature of the 2SE is its low-profile design. On modern cars, engine compartments are smaller and hood lines lower, so tall units are out of the question.

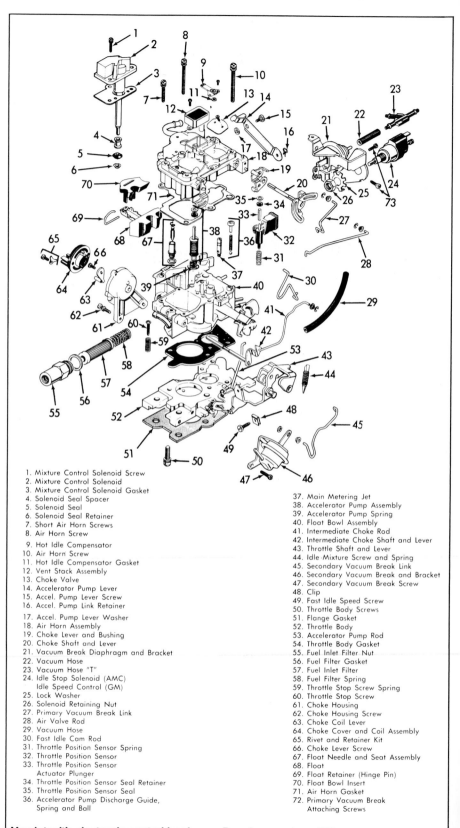

1. Mixture Control Solenoid Screw
2. Mixture Control Solenoid
3. Mixture Control Solenoid Gasket
4. Solenoid Seal Spacer
5. Solenoid Seal
6. Solenoid Seal Retainer
7. Short Air Horn Screws
8. Air Horn Screw
9. Hot Idle Compensator
10. Air Horn Screw
11. Hot Idle Compensator Gasket
12. Vent Stack Assembly
13. Choke Valve
14. Accelerator Pump Lever
15. Accel. Pump Lever Screw
16. Accel. Pump Link Retainer
17. Accel. Pump Lever Washer
18. Air Horn Assembly
19. Choke Lever and Bushing
20. Choke Shaft and Lever
21. Vacuum Break Diaphragm and Bracket
22. Vacuum Hose
23. Vacuum Hose "T"
24. Idle Stop Solenoid (AMC) Idle Speed Control (GM)
25. Lock Washer
26. Solenoid Retaining Nut
27. Primary Vacuum Break Link
28. Air Valve Rod
29. Vacuum Hose
30. Fast Idle Cam Rod
31. Throttle Position Sensor Spring
32. Throttle Position Sensor
33. Throttle Position Sensor Actuator Plunger
34. Throttle Position Sensor Seal Retainer
35. Throttle Position Sensor Seal
36. Accelerator Pump Discharge Guide, Spring and Ball

37. Main Metering Jet
38. Accelerator Pump Assembly
39. Accelerator Pump Spring
40. Float Bowl Assembly
41. Intermediate Choke Rod
42. Intermediate Choke Shaft and Lever
43. Throttle Shaft and Lever
44. Idle Mixture Screw and Spring
45. Secondary Vacuum Break Link
46. Secondary Vacuum Break and Bracket
47. Secondary Vacuum Break Screw
48. Clip
49. Fast Idle Speed Screw
50. Throttle Body Screws
51. Flange Gasket
52. Throttle Body
53. Accelerator Pump Rod
54. Throttle Body Gasket
55. Fuel Inlet Filter Nut
56. Fuel Filter Gasket
57. Fuel Inlet Filter
58. Fuel Filter Spring
59. Throttle Stop Screw Spring
60. Throttle Stop Screw
61. Choke Housing
62. Choke Housing Screw
63. Choke Coil Lever
64. Choke Cover and Coil Assembly
65. Rivet and Retainer Kit
66. Choke Lever Screw
67. Float Needle and Seat Assembly
68. Float
69. Float Retainer (Hinge Pin)
70. Float Bowl Insert
71. Air Horn Gasket
72. Primary Vacuum Break Attaching Screws

Varajet with electronic-control hardware. Drawing courtesy GM.

Quadrajet Design

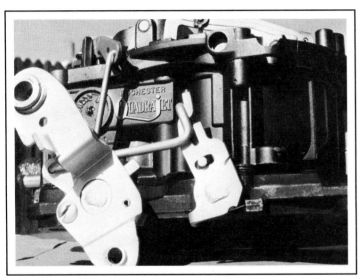

4

First-design Q-jet was revolutionary GM induction system.

1974 Chevrolet Q-jet with idle-stop solenoid.

HISTORY

The Q-jet is the only GM carburetor from the '60s to withstand emission-control requirements into the '80s. This chapter describes the Q-jet's design and examines system changes made between 1975 and 1986—including computer-controlled units.

From its original design up through the '70s, three models were built: The 4M—basic carburetor with manual choke, 4MV—automatic choke with manifold-mounted thermostatic choke coil, and 4MC—automatic choke with choke housing and thermostatic choke coil mounted on the float bowl. Except for the choke systems, all used the same operating principles until 1980. Computer-controlled units used on most passengers cars since 1981 perform the same functions, but have electronics controlling mixture and metering. For details, see pages 89—98 in the Quadrajet Service chapter.

In 1975, rigid emission standards were introduced that forced changes to idle, off-idle and main system metering control. The Adjustable Part Throttle (APT) system was placed in the fuel bowl for closer adjustment at the factory. These refined units were first used on passenger cars, and over the next six years, were fitted to trucks and commercial vehicles.

These carburetors can be identified by the distinctive shape of their air horn and the beginning digit of their part numbers—1 (one).

A nagging confusion was introduced with these external modifications. The primary metering rods with a 170 prefix part number are 0.080-in. shorter, and their part-throttle metering area—including tapers to the power-tip area—is quite different from pre-'75 designs.

43

The rods are not interchangeable. The confusion here is that either rod design interchanges with either type carburetor. Not only do they fit well, but without extremely careful inspection, they look alike. The part-number range of these primary rods is 17051336—52.

Also in 1975, an aneroid altitude compensator was introduced to automatically control the A/F ratio as the altitude increased. It never became popular—as ignition and emission hardware improved, manufacturers could meet standards without it.

In 1981, electronic metering control was introduced to meet the ever-stringent emission and economy regulations. Microprocessor control had a lasting effect on Q-jet service and modifications. See Electronic Carburetion in this chapter for design details.

Nearly every portion of the Q-jet has been subtly changed since its 1965 introduction, yet the *basic* design has not changed. The following text concentrates on explaining the Q-jet's fundamental carburetion systems. All in all, the original design concept was unusually good and was ahead of its time.

Q-JET FUEL CONTROL SYSTEMS

The Q-jet has always had two stages of operation. The primary has small bores with triple venturis. These establish stable and precise fuel control at idle and partial throttle. Until 1980, all primary fuel metering was done with tapered metering rods positioned in metering jets by a power piston that responded to manifold vacuum.

The secondary side has two large bores with increased air capacity to meet heavy throttle demands. An air valve controls the metering in the secondaries. It positions tapered metering rods in orifice plates to control fuel flow from the secondary nozzles in direct proportion to airflow through the secondary bores.

The central fuel reservoir prevents sloshing problems that cause cut-out and delayed fuel flow to the bores during abrupt maneuvers. The float system has a single float pontoon for ease in servicing.

Early units have a pressure-balanced fuel inlet valve to overcome problems with high fuel-pump pressures, and to allow a small float to control fuel through the large needle seat. It has a synthetic tip designed to reduce flooding problems caused by dirt clogging the inlet. Service problems in the field, and perhaps pressure from skeptics, caused RPD to return to a conventional needle and seat in 1967 models.

Early Q-jets have a sintered-bronze fuel filter mounted in the fuel-inlet casting, which is an integral part of the float bowl. The filter is easily removed for cleaning or replacement. It removes particles as small as 80 microns (0.003 in.) in diameter. These filters were later replaced by more efficient paper filters.

The primary side has six fuel-control systems: float, idle, main metering, power, accelerator-pump and choke. The following text describes all fuel-control systems in detail.

FLOAT SYSTEM

The Q-jet's central float reservoir, with its single pontoon float and fuel inlet valve, is unique. The fuel bowl is centered between the primary bores and adjacent to the secondary bores. This design assures adequate fuel delivery to all bores. The solid float pontoon is made of a closed-cell plastic material. Because it is more bouyant than a brass pontoon, a smaller float can maintain constant fuel levels.

A plastic filler block in the top of the float chamber just above the float valve prevents fuel slosh during severe braking. The filler maintains a more constant fuel level to prevent stalling during this action. It also reduces total fuel capacity, thereby reducing fuel vapor to the atmosphere.

Venting—The original float chamber was vented internally and externally. Internal vent tubes are in the fore and aft sides of the primary bore section, just above the float chamber. The elongated slots for the secondary metering rods also are a sizable vent.

The internal vent balances incoming air pressure beneath the air cleaner with air pressure acting on fuel in the bowl. Therefore, a balanced A/F mixture ratio can be maintained during part-throttle and power operations. The internal vent tubes let fuel vapor escape the float chamber during hot-engine operation. This prevents fuel vaporization from causing pressure buildup in the float bowl, which can cause fuel spilling into the throttle bores.

Early Q-jets have an external idle-vent valve that vents fuel vapors from the float bowl to the atmosphere during hot-engine soaking. The vent connects to the throttle and only opens at curb idle. It closes as the throttle is opened farther.

A temperature-controlled vent valve is used on some pre-emission Q-jets. A heat-

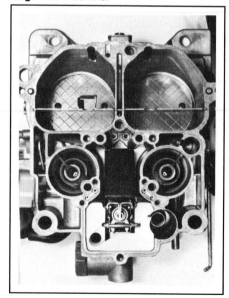

Top view of first-design Q-jet. Note small, very efficient primary venturis and big, big secondaries. Small central float bowl reduced fuel loss by evaporation during hot engine shutdowns.

Close-up of primary venturi/nozzle.

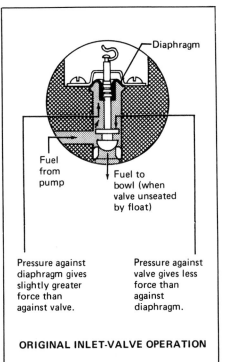

ORIGINAL INLET-VALVE OPERATION

Diaphragm

Fuel from pump

Fuel to bowl (when valve unseated by float)

Pressure against diaphragm gives slightly greater force than against valve.

Pressure against valve gives less force than against diaphragm.

Early pressure-balanced fuel inlet valve was replaced by conventional needle and seat in 1967.

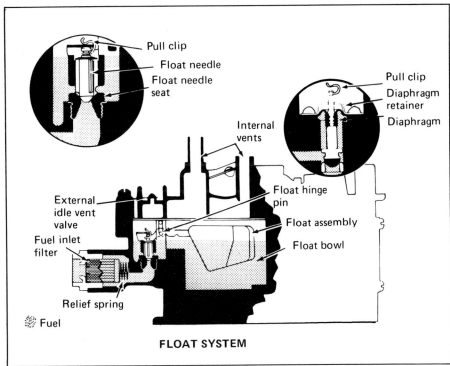

Pull clip

Float needle

Float needle seat

Internal vents

Pull clip

Diaphragm retainer

Diaphragm

External idle vent valve

Fuel inlet filter

Relief spring

Float hinge pin

Float assembly

Float bowl

Fuel

FLOAT SYSTEM

Basic float system from early Q-jet.

Primary metering jets (A) and secondary metering jets (B) are placed in central fuel bowl. Idle/off-idle fuel is metered from primary jet wells via idle tubes (C). Fuel wells for secondary side are at (D).

Plastic filler displaces void in top-forward portion of bowl, preventing fuel from rushing to area during braking.

Prominent internal vent is forward vertical stack (A) shown on first-design Q-jet. Most Q-jets from 1965—70 had external vent: it opened during closed throttle by tang off pump lever (B).

sensitive bimetal strip holds the vent valve in position beneath the idle-vent valve arm. The valve is held on its seat during modest operating temperatures. Closing the valve keeps fuel vapor in the carburetor to supply extra fuel for easy cold-engine starting. At higher underhood temperatures, the bimetal unseats the valve so fuel vapor escapes, thus improving hot-engine starting and idling at these temperatures.

During hot-engine operation, except during idle, the valve is closed by a spring-steel vent arm operated by a wire lever on the pump lever. Opening the throttle from idle closes the vent valve, regardless of underhood temperature.

Some Q-jets have a vacuum vent switch in the air horn to vent vapor to the air cleaner during engine operation and to a charcoal canister when the engine is off. A spring beneath the vacuum-diaphragm piston forces it upward to close the air-cleaner vent and open the canister vent. During heavy acceleration or low manifold vacuum, the air-cleaner vent is held open by an actuating arm on the pump lever. This maintains internal carburetor pressure balance at all times when the engine is running.

Newly designed systems have since eliminated this means of venting during hot soak because of vapor loss to the atmosphere. Current systems pipe fuel vapor to a charcoal-filled canister where it is retained until the engine is run. Running purges the vapor and routes it to the combustion chamber.

IDLE SYSTEM

The Q-jet has an idle system in its primaries that supplies correct mixture during idle and off-idle operation. It operates during these conditions because insufficient air flows through the venturis to get correct metering from the main discharge nozzles.

The idle system is only in the two primary bores. Each bore has an idle tube, idle passage, idle-air bleed, idle-channel restriction, idle-mixture adjustment needle, curb-idle discharge hole and off-idle discharge port.

During curb-idle, the throttle valve is held slightly open by the idle-speed adjusting screw. The small amount of air passing between the primary throttle valve and bore is regulated by this screw to get idle speed. Fuel is added to the air by applying intake-

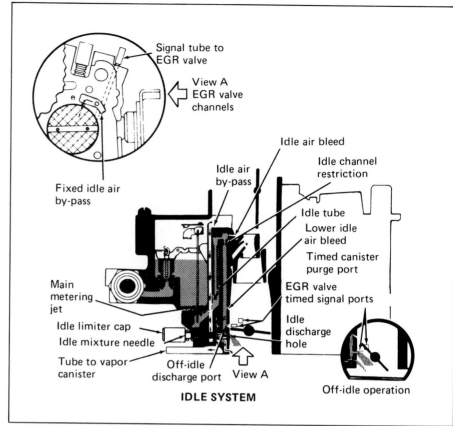

IDLE SYSTEM

Early Q-jet idle system.

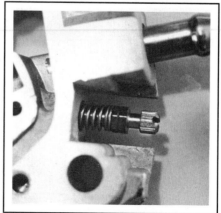

Each primary bore of non-electronic Q-jet has one idle-mixture adjustment screw.

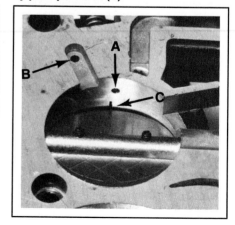

Curb-idle adjustment feed orifice (A) and part of off-idle feed slot (B) are shown with throttle blade slightly open. Fixed idle-air by-pass port is at (C).

46

manifold vacuum—low pressure—to the idle-discharge hole below the throttle valve.

Fuel travels a complex path in the idle system. It is drawn from the float bowl through the primary main metering jets into the main fuel well. Then, it passes from the main fuel well into the idle passage, where it is picked up by the idle tube. Fuel metered at the idle-tube tip goes up the idle tube. At the top of each idle tube, the fuel is mixed with air supplied through an idle-air bleed. The A/F mixture crosses over to the idle-down channel, where it passes through a calibrated idle-channel restriction.

Next, it follows the idle channel past the lower idle-air bleed hole and off-idle discharge port, just above the primary throttle valve, where it is mixed with more air. The mixture continues to the curb-idle discharge hole, enters the carburetor bore, and finally mixes with air passing around the slightly open throttle valve. It then enters the intake manifold and is conducted to the engine cylinders as a combustible mixture.

The adjustable idle-mixture needles control the amount of fuel mixed with the air going to the engine. Turning the mixture screw in (clockwise) decreases the fuel discharge to give a leaner idle mixture. Turning the mixture screw out (counterclockwise) increases fuel discharge to richen the idle mixture.

The idle-air bleed is above each idle tube. This bleed also serves as a siphon break and a calibration aid. The air bleed leads from the top of this channel into the air horn above the venturi. It aids the escape of fuel vapor in the idle-tube channel during hot-engine idling and prevents it from mixing with fuel being picked up by the idle tube. This gives a more consistent idle-fuel mixture during hot-engine idling.

Some Q-jets have a fixed idle-air-bypass system. An air channel leads from the top of each primary bore in the air horn to a calibrated hole below each primary throttle, or to the throttle-body gasket surface. At normal idle, extra air passing through these channels supplements the air passing by the slightly-opened primary throttles.

This fixed air supply reduces the air requirement through the throttles, so they can be nearly closed at idle. The amount of air flowing through the carburetor venturis is less, so the main nozzles will not *feed* (drip) at idle. Because the venturi system in the primary bores is so efficient, this idle-

Q-jet throttle body and main body casting passages: idle by-pass (A) delivers air from above venturi through fixed orifices in throttle body (not all models), idle-down channel (B), timed spark (horizontal slot above primary throttle) (C), manifold-vacuum port (D), and power system manifold-vacuum port (E).

Arrows (left photo) indicate bushed air bleeds, knows as "small-bleed" type. Arrows (right photo) show "large-bleed" type used on some Chevrolet models.

air-bypass system is used on some applications where large amounts of air are required to maintain the idle speed. For example, it is especially helpful when a high-performance cam is used because more airflow is required to maintain acceptable idle speed.

On emission-controlled applications, the idle-mixture discharge holes are reduced to prevent a too-rich idle adjustment (rich roll) if the mixture needles are turned too far out. On some models, the size of the idle-discharge holes was increased when idle-air bleed sizes were enlarged to combat fuel percolation problems at high-temperature operation.

Another feature sometimes added to emission-controlled carburetors is an adjustable idle-air bleed system. A separate air channel leads from the top of the air horn to the idle-mixture cross channel. A tapered-head adjustment screw at the top of the channel is used to control the amount of air bleeding into the idle system. This bleed is factory-adjusted when the carburetor is flow-checked, and a triangular spring clamp is forced over the vent tube to discourage field adjustments.

OFF-IDLE OPERATION

When primary throttle valves are opened from curb idle to increase engine speed, additional fuel is needed to combine with the extra air entering the engine. This fuel is supplied by the slotted off-idle discharge ports. As the primary throttle valves first open, they pass by the off-idle ports, and gradually expose them to high engine vacuum below the throttle valves.

Without correct fuel-flow response to engine demand, sag and backfires happen when tipping in the throttle from idle. The accelerator pump helps unless throttle movement is very slow. Opening the throttle valves sufficiently increases air velocity through the venturi to cause low pressure at the lower idle-air bleeds. As a result, fuel begins to discharge from the lower idle-air bleed holes and continues to do so throughout part-throttle operation to WOT. At higher airflows (heavy throttle) the idle and off-idle system supplies only a small portion of the needed A/F mixture.

HOT-IDLE COMPENSATOR

The hot-idle compensator is in a chamber at the rear of the float bowl, adjacent to the secondary bores, or in the primary side

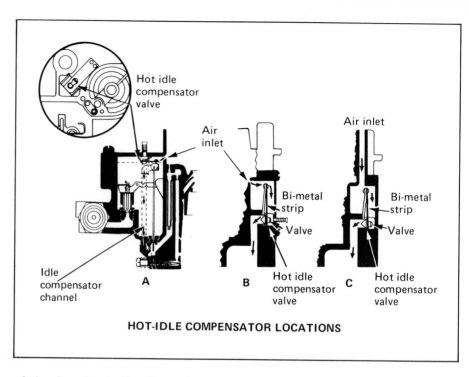

HOT-IDLE COMPENSATOR LOCATIONS

of the float bowl. It offsets richening effects caused by excessive fuel vapor during hot-engine operation. Hot-idle compensators were used on Q-jets from 1967 to the late '70s.

The compensator is a thermostatically controlled valve made of a heat-sensitive bimetal strip, a valve holder and bracket. The valve closes off an air channel leading from a vent in the air horn to a point below the secondary throttle valves.

The compensator valve is held closed by the bimetal strip's tension. During extreme hot-engine operation, excessive fuel vapor entering the engine manifold causes too rich a mixture, rough idle and stalling. At a predetermined temperature, when extra air is needed to offset the richening effects of fuel vapor, the bimetal strip bends to unseat the compensator valve. The air channel opens, air is drawn into the engine manifold to offset the richer mixture, and smooth idle results. When the engine cools and the extra air is not needed, the bimetal strip closes the valve and operation returns to standard mixtures.

MAIN METERING SYSTEM

The main metering system supplies fuel from off-idle to WOT operation. Q-jets have two small primary bores that feed fuel and air throughout this range, metering fuel accordingly. Multiple venturis are used for

fine, stable metering control during light engine loads.

The system works when airflow through the venturi is high enough to maintain efficient fuel flow from the main fuel-discharge nozzles. The main system begins to feed fuel when the idle system can't meet the engine's mixture demands.

The main metering system consists of main metering jets, vacuum-operated metering rods or solenoid-controlled

Hot-idle-compensation assembly is usually mounted in recessed area at rear of carburetor bowl. Some HIC valves are in primary side of float bowl.

metering rods—the latter are explained on page 59—a main fuel well, main-well air bleeds, fuel-discharge nozzles and triple venturis. The system operates as follows:

During cruising and light engine loads, engine manifold vacuum is high. Manifold vacuum holds the part-throttle portion of the main-metering rods down in the main-metering jets against spring tension. Vacuum is supplied to a vacuum-operated piston through an orifice in the throttle body. This piston is called the *power piston*, but it also controls the position of the metering rods at part-throttle. Fuel flow from the bowl is metered between the metering rods and the main-jet orifice.

Fuel Flow—When the primary throttle valves are opened beyond off-idle, air velocity increases in the carburetor venturi. Fuel flows from the float bowl through the main metering jets into the main fuel well and is mixed with air from the bleed vent at the top of the main well and side bleeds. Fuel in the main well is then mixed with air from the main-well air bleeds and passed through the main discharge nozzle into the boost venturi. At the boost venturi, fuel mixture combines with air entering the throttle bores, passes through the intake manifold, and is delivered to the cylinders as a combustible mixture.

The main metering system is calibrated by air bleeds and by tapered and stepped metering rods in the main jets. A vacuum-responsive power piston positions the metering rods in the jets to supply fuel as required by various engine loads. This is described in the following section on the power system.

Three types of primary main metering rods are used in non-electronic Q-jets. The 1967 and earlier models use rods with a single taper at the metering tip. The 1968 and later models use rods with a double taper at the metering tip, identified with the suffix B after the part number. The third type was used in 1975—1980 vehicles and is indicated with part-number prefix 170. All the part numbers identify the diameter of a specific point on the part-throttle portion of the rods.

Adjustable Part-Throttle (APT)—The accompanying photo illustrates the early Q-jet design. The screw behind plug (C) is factory-adjusted to position the power piston during high part-throttle vacuum. On many units the piston/metering rods were adjusted so high the vehicle was actually

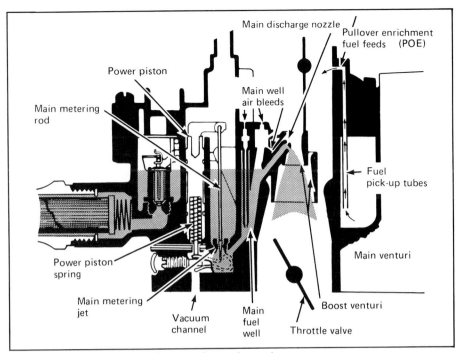

Q-jet main metering system for non-electronic versions.

running on the beginning of the taper. This caused premature richness whenever the throttle was modestly applied. Metering rods used with the APT are stamped with the suffix B after their identifying number.

The APT helps part-throttle performance because all of the fuel at part-throttle is supplied through the main metering jets.

Pull-Over Enrichment—Some Q-jets are equipped with a fuel pull-over enrichment

circuit to supplement fuel feed from the primary main-discharge nozzles. These supplementary fuel feeds allow lean A/F mixtures during part-throttle, while supplying extra fuel for good performance at higher speeds.

This circuit consists of two calibrated holes, one in each primary bore, just *above* or *below* the choke valve, and located above the venturis. During high airflows, low pressure in the air horn pulls fuel from

Vacuum piston supporting primary metering rod hanger is in cylinder protruding from fuel-bowl floor. Tapered portion of metering rods is in orifice of metering jets.

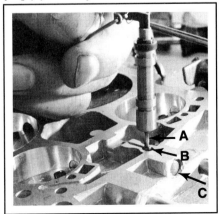

APT system raises power piston so primary metering rods are precisely positioned in main jets to supply required part-throttle fuel mixture. Adjustable lever at (A), barely visible, is stop for power-piston tip (B). Adjusting screw for APT lever is hidden by plug (C) after adjustment at RPD.

these holes. The system starts feeding fuel at approximately 8 pounds of air per minute and continues to feed at higher rates.

If the calibrated holes are *below* the choke valve, the circuit feeds additional fuel when the choke is closed, for better cold-engine starting. Calibrated air bleeds in the air horn are used with this system, called a *choke enrichment* system. It is described in the Choke section later in this chapter.

POWER SYSTEM

This supplies extra mixture enrichment for power requirements during acceleration or high-speed operation. The richer mixture enters through the main metering system in the primaries and secondaries.

The power system in the primary side is actuated by a vacuum piston and spring in a cylinder connected by a passage to intake-manifold vacuum. The spring beneath the vacuum-operated power piston tends to push the piston up. Manifold vacuum tries to pull it down.

During part throttle and cruising, high manifold vacuum holds the power piston down against spring tension—the larger diameter of the main-metering-rod tips is kept in the main-metering-jet orifices. A rather lean mixture results for best fuel economy. Mixture enrichment is not necessary at this point.

When engine load increases, manifold vacuum decreases and spring tension raises the power piston and metering rods. The tapered primary-metering-rod section lifts out of the main-metering-jet orifices—a smaller-diameter portion of the tip is in the jet. So, more fuel passes through the main metering jets to richen the mixture flowing into the primary main wells and out the main discharge nozzles.

On some Q-jets, two springs are used beneath the power piston. The smaller (primary) spring seats in the center of the piston. A larger diameter (secondary) spring surrounds the smaller spring. It exerts added force upward on the power piston. It starts the first stage of power enrichment at 8—10-in.Hg manifold vacuum. The center spring gradually continues enrichment until it is completed at 5—7-in.Hg manifold vacuum.

SECONDARY SYSTEM

As engine speed increases, the primaries can't meet the engine's airflow demands and the secondaries begin to operate. The secondary section contains:
- Throttle valves.
- Spring-loaded air valves.
- Metering orifice plates.
- Secondary metering rods.
- Main fuel wells with air-bleed tubes.
- Fuel-discharge nozzles.

- Accelerating wells and tubes.

The secondary side operates as follows.

Fuel Flow—When engine demand requires more A/F than the primary bores can supply, the primary-throttle lever opens the secondary throttle valves with its connecting linkage to the secondary throttle-shaft lever. As airflow through the secondary bores creates a low pressure (vacuum) beneath the air valve, atmospheric pressure on top of the air valve forces it open against spring tension. The required air for increased engine speed then flows past the air valve.

As the air valve begins to open, its upper edge passes the accelerating-well ports, exposing them to manifold vacuum. The ports immediately feed fuel from the wells and continue to do so until they are depleted. The accelerating ports prevent momentary leanness as the valve opens and the secondary nozzles begin to feed fuel. Some Q-jets do not have accelerating wells.

The secondary main-discharge nozzles, one for each secondary bore, are just below the air valve and above the secondary throttle valves. They feed fuel as follows.

When the air valve opens, it rotates a plastic cam attached to the main air-valve shaft. The cam lifts a lever attached to the secondary main metering rods and pulls the rods out of the secondary-orifice plates (jets). Fuel from the float chamber flows through the secondary-orifice plates into secondary main wells, where it is mixed with air from the main-well tubes. The air-emulsified fuel mixture travels from the main wells to the secondary-discharge nozzles and into the secondary bores. Here, the fuel mixture is combined with air passing through the secondaries to supplement the A/F mixture delivered from the primaries. It is then distributed into the manifold and cylinders as a combustible mixture.

As the throttle valves are opened farther, airflow through the secondaries opens the air valve more, which in turn, lifts the secondary metering rods farther out of the orifice jets. The metering rods are tapered so that fuel flow through the jets is directly proportional to airflow through the secondary bores. Consequently, correct A/F mixtures supplied by the secondary bores are governed by the depth of the metering rods in the orifice plates in relation to air-valve position. This relationship is factory-adjusted to meet the air/fuel requirements

Some Q-jets have pull-over enrichment tubes picking up fuel directly from bowl (left photo, white arrows). These connect with feed holes (right photo, black arrows) in air horn. When velocity increases in primary side (high speed or medium-to-heavy acceleration), depression created at holes pulls fuel into primaries. This allows extremely lean metering at low speeds to pass emission-certification requirements, yet provides adequate fuel for good high-speed performance. Wide boss around center bowl vent indicates carburetor equipped with pull-over circuit. It is also called *high-speed fuel feed.*

for a specific engine model. No further adjustment should be required for normal applications.

When the air valve has lifted the secondary metering rods to their maximum height, the smallest-diameter tips of the rods are in the orifice jets, supplying the richest full-power mixture for high-rpm driving. Complete specifications for secondary metering rods are listed on page 152.

The secondaries also have these design features:

● Main-well bleed tubes extend below the fuel level in the well. These bleed air into the well to emulsify the fuel with air and improve atomization as the mixture leaves the secondary-discharge nozzles.

● Many secondary metering rods through 1972 had a slot milled in their side to ensure adequate fuel in the secondary fuel wells. These slots were thought necessary because the rods fitted so closely in the secondary metering orifice plates (jets) when the air valve was closed.

During hot-engine idle or hot-soak, the fuel could boil away in the fuel well and leave it empty. The slots let fuel bypass the orifice jets so the main wells were filled during these conditions. This ensured immediate fuel delivery from the secondary fuel wells. In 1973 the slots were eliminated by setting an extra 0.001-in. clearance between the rod and orifice jet.

● A directional baffle plate in each secondary bore extends up and around the secondary fuel-discharge nozzles. These baffles aid in directing the airflow for improving distribution in the manifold.

● A vertical air-horn baffle is used on some models to prevent incoming air from forcing fuel into the secondary wells through the bleed tubes. This prevents secondary-nozzle lag on heavy acceleration.

AIR-VALVE DASHPOT

The secondary air valve in early Q-jets has an attached piston acting as a damper to prevent valve oscillation caused by engine pulsations. The damper piston operates in a well filled with fuel from the bowl. Piston motion is retarded by fuel bypassing the piston when it moves up in the fuel well. Piston drag slows air-valve opening and prevents lag in the secondary discharge nozzles.

All Q-jets made since 1967 have an air-valve dashpot operated by the choke

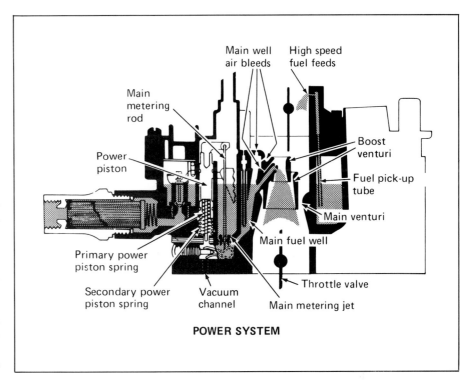

POWER SYSTEM

Baffle in secondary side channels air past nozzles until air valve is fully open. This creates low-pressure and causes fuel to flow from nozzles.

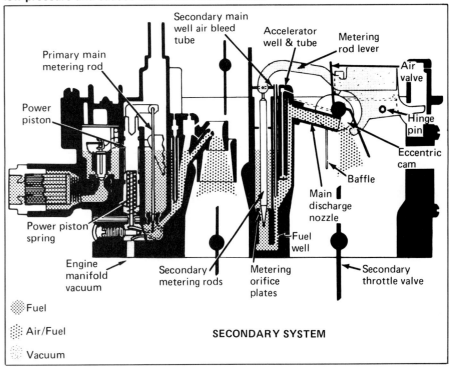

SECONDARY SYSTEM

As secondary air valve opens, forward edge of blades passes secondary fuel discharge ports, exposing them to reduced pressure. This pulls shot of fuel that eliminates lean sag as air valve starts to open. A few Q-jets models didn't have these ports.

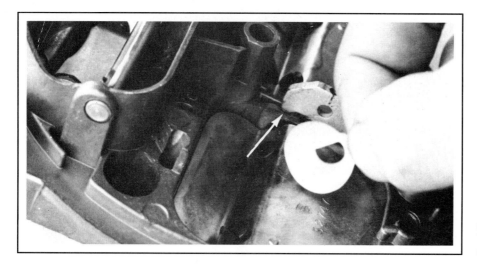

Spare plastic cam is held to show its shape. These cams can sometimes crack or wear and fail.

Metal pick-up lever between air valves rides on cam (arrow). Secondary metering rod hanger mounts to center lever. When air valve opens, high part of cam lifts lever and secondary metering rods rise for automatic enrichment at high airflows.

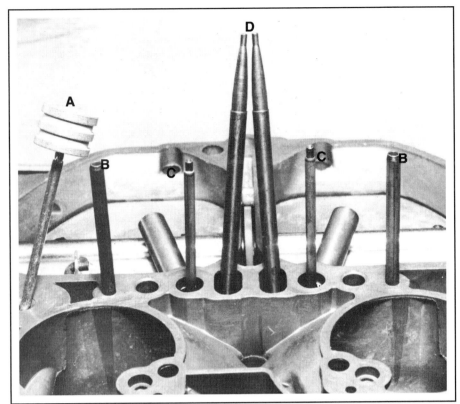

Early-model air horn lifted off and inverted reveals: (A) air valve damping piston (some models), (B) fuel feed tubes for secondary accelerating well discharge holes, (C) secondary main well bleed tubes and (D) secondary metering rods.

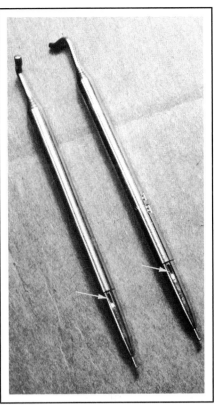

Secondary rods are precision pieces. Arrows point out fuel slots referred to in text. They are expensive and some sizes are hard to replace.

vacuum-break diaphragm unit. The air valve is connected to the vacuum break by a rod. Whenever manifold vacuum is above 5 in.Hg, the vacuum-break diaphragm is seated. The valve plunger is then held fully in against spring tension. The vacuum-break rod is moved to the front end of the slot in the air-valve lever, and the air valve closes.

During acceleration, or during loads when the secondary throttle valves are opened, manifold vacuum drops. A spring in the vacuum-break diaphragm overcomes vacuum pull and forces the plunger and link out. This plunger, link movement and air-valve-opening rate are controlled by a calibrated restriction in the vacuum inlet to the diaphragm. This restriction damps the valve opening so fuel flow can begin at the secondary discharge nozzles.

ACCELERATOR-PUMP SYSTEM

When the throttles are opened rapidly for acceleration, airflow and manifold vacuum change almost instantaneously. Because fuel is heavy and lags behind airflow, a momentary leaness results. The accelerator pump supplies extra fuel for smooth operation during this condition.

The accelerator-pump system is located in the primaries. It is a spring-loaded pump plunger and pump-return spring operating in the pump well. The pump plunger is controlled by a lever connected to the throttle lever by a rod.

When the plunger moves up in the pump well during throttle closing, fuel from the float bowl enters the well through a slot in its top. The floating pump cup moves up and down on the pump-plunger head. When the pump plunger is moved up, the

flat on the top of the cup unseats from the flat on the plunger head. Fuel moves through the inside of the cup into the bottom of the pump well.

This floating cup also vents any fuel vapor in the well—so a charge of liquid fuel is kept in the fuel well beneath the pump-plunger head.

When the primary throttle valves are opened, connecting linkage forces the pump plunger down. The pump cup seats instantly and fuel is forced through the pump-discharge passage. There it unseats a check ball and passes on to pump jets in the air horn. Here, fuel sprays into the venturis of the primaries. The pump plunger is spring-loaded. The upper-duration spring is balanced with the bottom pump-return spring so a sustained charge of fuel is delivered during acceleration.

The pump-discharge check ball seats in the pump-discharge passage when the plunger moves up so air will not draw into the passage. Otherwise, a momentary acceleration lag would result.

During high-speed operation, vacuum exists at the pump jets. A cavity beyond the jets is vented to the top of the air horn, outside the carburetor bores. This vent acts as a suction breaker—when the pump isn't in use, fuel won't be pulled out of the pump jets into the venturi.

CHOKE SYSTEM

There are many variations of design used to set the choke, modulate it according to engine load and temperature, and release it when the engine warms up. Some of these design variations are complex. Study the accompanying drawings. Other versions will probably follow because releasing the choke quickly is an important requirement for passing emissions tests when an engine must be cold-started.

The choke valve is on the primary side of the carburetor. It establishes correct A/F mixture enrichment for cold-engine starting and warmup. The air valve is locked closed until the engine is warm and the choke valve is completely open.

The choke system consists of:
- Choke valve in the primary side of the air horn.
- Vacuum-diaphragm unit.
- Fast-idle cam.
- Connecting linkage.
- Air-valve-lockout lever.
- Thermostatic coil.

The thermostatic coil is usually mounted on the engine manifold and connected to the intermediate choke-shaft-and-lever assembly by a *choke rod*. Choke operation is controlled by intake-manifold vacuum, the offset choke valve, temperature and throttle position.

The thermostatic coil in the intake manifold—or on the carburetor body in some applications—is calibrated to hold the choke valve closed when the engine is cold. When the choke valve is closed, the air-valve lockout lever is weighted so its tang catches the upper edge of the air valve and keeps it closed, too. On some models, the secondary throttles may be locked-out when the choke is closed. Both types of lockout are used in Q-jets to prevent secondary operation until the engine is warm.

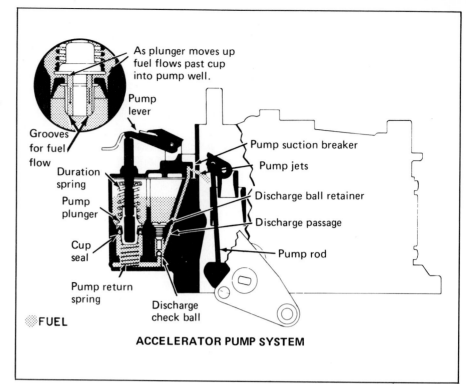

ACCELERATOR PUMP SYSTEM

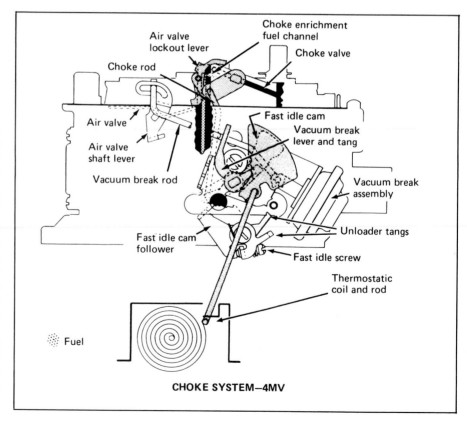

CHOKE SYSTEM—4MV

During cold-starting, air pressure against the top of the choke valve causes it to open slightly. Airflow is restricted and the mixture is richened. After starting, manifold vacuum to the vacuum-break diaphragm opens the choke valve so the engine will run without loading or stalling. As the throttle opens, the fast-idle-cam follower lever on the primary throttle shaft-end drops from the highest step on the fast-idle cam to the second step. This sets throttle and idle for running until the engine warms the thermostatic coil. As the coil becomes heated, its tension relaxes, and the choke valve opens farther. Opening continues until the coil is fully relaxed and the choke is wide open.

The vacuum-break modulating spring allows the vacuum break to vary choke position according to ambient temperature. This spring connects to the vacuum-break link and changes choke openings, depending on the closing force of the thermostatic coil. As this force increases, caused by cool temperatures, the link moves in the slotted lever until the modulating spring overcomes the coil force, or the link is in the end of the slot. So, less vacuum break results and lets the choke valve stay closed during cooler weather. During warmer weather, more vacuum break pulls the choke blade open.

Split-Choke—Early MV Models—This operates during the last few degrees of choke-thermostat rotation. It maintains fast-idle speed after the choke is essentially off, but the engine isn't fully warmed-up.

Its operation is controlled by a torsion spring on the intermediate choke-lever shaft. Air-pressure action on the offset choke valve forces it open against thermostatic coil tension. In the last few degrees of thermostatic-coil opening motion, a tang on the intermediate-choke lever contacts the end of the torsion spring.

The tang keeps the fast-idle-cam follower lever on the last step of the fast-idle cam, which maintains fast-idle until the engine is thoroughly warm. The spring restrains the thermostatic coil until the coil is hot enough to pull on the intermediate choke lever and overcome spring tension.

When the engine is warm, the choke coil pulls the intermediate choke lever completely down, which lets the fast-idle cam rotate so the cam follower disengages from the last step on the cam. Normal idle begins. When the choke-shaft lever moves up, the end of the rod strikes a tang on the air-valve lockout lever and pushes the lockout tang upward and frees the air valve.

Spring-Assist Choke Closing System—Some 4MV carburetors use a torsion assist spring on the intermediate choke shaft. The torsion spring assists in closing the choke valve to ensure good engine starting when the engine is cold. It exerts pressure on the vacuum-break lever to force the choke valve closed. Spring tension is overcome by the choke thermostatic coil when the engine warms up.

Other 4MV models have the choke-closing assist spring on the vacuum-break plunger stem—instead of on the intermediate choke shaft. The spring's tension is added to the closing tension of the remote choke thermostatic coil to close the choke valve during starting. When the engine starts and the choke vacuum-break diaphragm seats, the closing spring retainer hits a stop on the plunger stem and no longer exerts pressure on the choke valve.

With this design, the tang on the vacuum-break diaphragm plunger is removed to eliminate the fast-idle cam pull-off. Also, the slot for free travel of the air-valve dashpot link is on the vacuum-break plunger instead of on the air-valve shaft lever.

Delayed Vacuum-Break System—During cold starting, vacuum applied to the choke vacuum-break diaphragm opens the choke valve against thermostatic choke coil tension. An internal air-bleed check valve inside the diaphragm keeps the choke valve from opening too fast.

When the engine starts, vacuum acting on the internal valve bleeds air through a small hole in it, which allows the vacuum-diaphragm plunger to move in slowly. When the diaphragm is fully seated, which takes a few seconds, the choke valve will remain in the vacuum-break position until the engine begins to warm and relax the thermostatic coil.

Some models include a rubber-covered

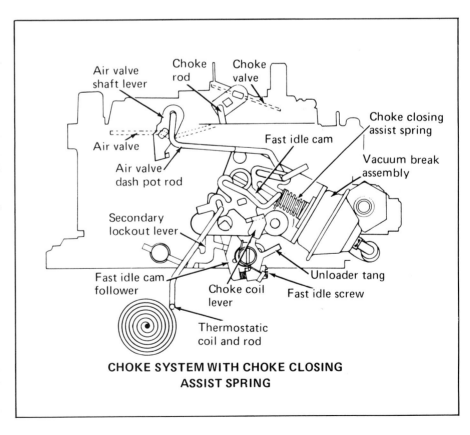

CHOKE SYSTEM WITH CHOKE CLOSING ASSIST SPRING

filter over the bleed hole in the vacuum tube to the rear vacuum break. The rubber-covered filter must be removed and the bleed hole closed with a piece of tape when making system adjustments. Besides the internal-bleed check valve, some car applications have a separate vacuum-delay tank. It is connected in series to a second vacuum tube on the vacuum diaphragm unit to further delay the choke vacuum-break-diaphragm operation.

The check valve in the vacuum diaphragm unit is designed to lift off its seat and allow the diaphragm plunger to extend out when spring force against the diaphragm is greater than vacuum pull. This movement adds enrichment as needed when accelerating a cold engine—the choke coil closes the choke slightly.

Some 4MV models use a calibrated restriction in the vacuum inlet to the vacuum-break diaphragm unit instead of an internal air-bleed check valve. The calibrated restriction also delays vacuum to the diaphragm unit to retard opening of the choke for good engine starting.

Bucking Spring—A spring-loaded plunger used in the vacuum break on some 4MV models offsets choke thermostatic coil tension and balances opening of the choke valve with choke-coil tension. This further refines A/F mixtures because the coil, which senses engine and ambient temperatures, allows the choke valve to gradually open against *bucking spring* tension in the diaphragm-plunger head. In cold temperatures, the extra tension of the thermostatic coil overcomes the diaphragm-plunger (bucking) spring. The choke-valve opening is less and mixture is richer. In warmer temperatures, the thermostatic coil has less tension and cannot compress the spring as much. More choke valve opening supplies leaner mixtures.

Dual-Delayed Vacuum-Break System—Other Q-jets have a front (main) vacuum-break diaphragm combined with another rear (auxiliary) break unit. This combination works as follows: During engine cranking, the choke valve is held closed by thermostatic coil tension—restricted airflow through the carburetor equals a richer starting mixture.

When the engine starts, manifold vacuum is applied to both vacuum-break diaphragm units mounted on the side of the float bowl. The *main* unit opens the choke valve so the engine will initially run without loading or stalling. As the engine is wetted and friction decreases after starting, an internal bleed in the *auxiliary* unit causes it to open the choke valve until the engine will run at a slightly leaner mixture.

The auxiliary unit has a spring-loaded plunger. It offsets choke thermostatic coil tension and balances the opening of the choke valve against choke coil tension. This enables further mixture refinement because the coil, which senses engine temperature, will gradually open the choke valve.

Early 4MC Models—These have a choke mechanism that includes an adjustable vacuum-break diaphragm. They operate like 4MV models, with one exception.

When the engine is running, manifold vacuum is applied to the vacuum diaphragm. An adjustable plastic plunger is on this diaphragm. Engine vacuum pulls in on the diaphragm and the plunger strikes the vacuum-break tang inside the choke housing. This action rotates the intermediate choke shaft, and via connecting linkage, opens the choke valve.

Later 4MC Models—These use a delayed vacuum break. An internal bleed check valve in the vacuum-break diaphragm delays diaphragm action a few seconds. The delay permits the manifold to be wetted and engine friction to decrease. When the vacuum-break point is reached, the engine will run without loading or stalling.

As the choke valve moves to the vacuum-break position, the fast-idle-cam follower lever drops from the highest step to the next lower step when the throttle is opened. The engine now has sufficient fast-idle speed and correct fuel mixture for running until it is warm enough to heat the thermostatic coil in the choke housing.

Engine vacuum pulls heat from the manifold heat stove into the choke housing and gradually relaxes choke-coil tension. Choke valve opening continues until the thermostatic coil is completely relaxed, the choke valve is wide open, and the engine is warm.

Vacuum Reindexing Unit—Beginning in 1974, Cadillac models have a vacuum-operated diaphragm plunger on the remote choke coil assembly to improve cold-engine starting. The assembly is called a *vacuum reindexing unit*. It applies additional pressure on the choke coil, which increases the pressure holding the choke valve closed when starting the engine. This

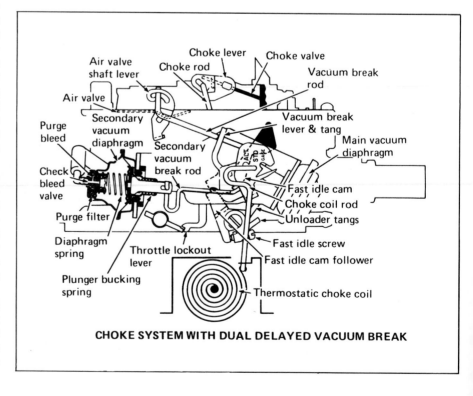

CHOKE SYSTEM WITH DUAL DELAYED VACUUM BREAK

design supplies correct A/F mixtures for engine starting, yet the choke opens quickly for economy and reduced emissions.

The vacuum reindexing unit is a spring-loaded diaphragm operated by direct manifold vacuum. When the engine is not running, no vacuum is applied to it. The spring on top of the diaphragm pushes it and the plunger down. The plunger end strikes a lever attached to the center choke-coil shaft, rotates the choke coil counterclockwise, and increases pressure on the thermostatic coil. This, in turn, increases the pressure holding the choke valve closed for engine starting.

After the engine starts, manifold vacuum acts on the vacuum diaphragm top and pulls the diaphragm plunger up against spring tension until it clears the lever on the remote choke-coil shaft. The choke coil then operates normally—it gradually opens as the thermostatic coil warms up.

A bleed check valve in the diaphragm's top retards the plunger's upward movement to gradually reduce choke-coil pressure so the engine will not stall. A purge hole in the vacuum tube draws filtered air into the vacuum channel. The air removes engine vapors or dirt that could plug the internal filter or bleed hole in the check valve.

Check the unit's operation by making sure the plunger is fully extended with the engine off; it should slowly move to the full-up position when the engine starts and runs. If it doesn't work, replace it.

Unloader—All Q-jet choke systems have an unloader mechanism that partially opens the choke valve should the engine become flooded when starting. To unload the engine, fully depress the accelerator—to the floorboard—so the throttle valves are wide open. A tang on a lever on the choke side of the primary throttle shaft contacts the fast-idle cam and forces the choke valve open. Extra air enters the bores and leans out the mixture so the engine will start.

Choke Enrichment System—Some Q-jet models use a system that supplements fuel from the primary main discharge nozzles for cold-engine starts. Two calibrated holes, one in each primary bore, are in the air horn just *below* the choke valve to supply fuel for enrichment during cold engine cranking. The extra fuel comes from channels to the secondary accelerating-well pickup tubes.

During warm-engine operation, the two calibrated holes in the air horn also supply

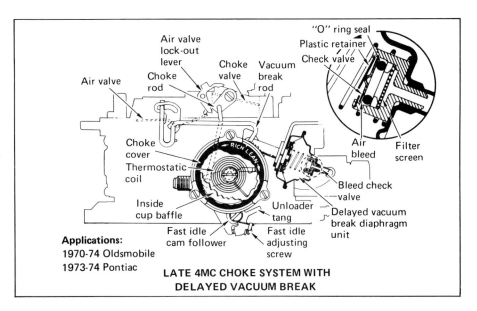

Applications:
1970-74 Oldsmobile
1973-74 Pontiac

LATE 4MC CHOKE SYSTEM WITH DELAYED VACUUM BREAK

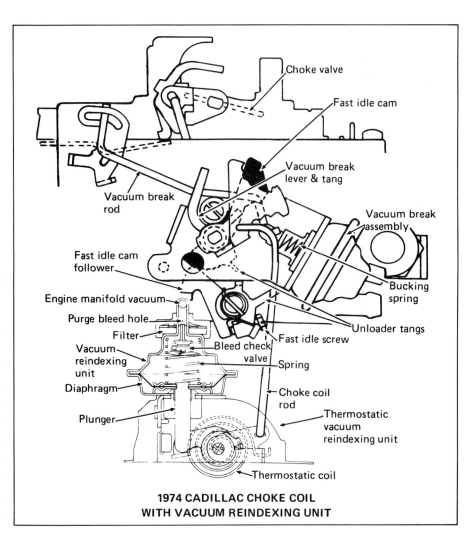

1974 CADILLAC CHOKE COIL WITH VACUUM REINDEXING UNIT

pull-over enrichment. They feed extra fuel at higher airflows—faster engine speeds—to supplement fuel flow in the primary bores.

Other Q-jets have a fuel pull-over enrichment method similar to this choke enrichment, except the two calibrated holes, one in each primary bore, are just *above* the choke valve. They supply extra fuel *only* during higher airflows. They don't feed fuel at closed-choke during engine cranking.

MICROPROCESSOR CONTROL OF AN ENGINE

The Electronic Control Module (ECM) is the microprocessor "brain" of the Computer Command Control (CCC) system on GM cars. It accepts inputs from sensors on the vehicle, processes this data, and transmits outputs to control engine operation. It has a memory for storing data about engine malfunctions that have happened or are happening.

When the ECM has identified a problem, CHECK ENGINE lights up on the dashboard. The problem has been logged in memory as a *trouble code.* Codes are retrieved from memory with a scanner tool plugged in the Assembly Line Diagnostic Link (ALDL) connector, and then interpreted. Factory shop manuals list which code indicates what sensor problem, and describe diagnostic procedures.

Remember, the ECM is the controlling microprocessor. To make decisions, it needs inputs from different sensors throughout the vehicle. If one of these is faulty, the ECM may send incorrect signals and cause the engine to run poorly.

Sensors supply input data to the ECM for these parameters:

- Coolant temperature.
- Vehicle speed.
- Manifold vacuum.
- Oxygen content in exhaust.
- Engine speed.
- Barometric pressure.
- Detonation (pinging).

After reading this data, the ECM processes it and adjusts or controls:

- Early fuel evaporation (heater in carburetor base).
- Mixture Control (M/C) solenoid.
- Engine spark timing.
- Exhaust Gas Recirculation (EGR).
- Charcoal canister purging.
- Air conditioning clutch.
- Idle speed.
- Transmission torque-converter lockup.

These are the basic sensor inputs to the ECM and its outputs. Different vehicle models and engines may have other combinations of sensors and outputs; see a factory shop manual for your particular application.

E4ME series carburetor. Note M/C solenoid and TPS connectors, and ISC. Electronic control of metering replaces vacuum/mechanical regulation. Photo courtesy Rochester Products Division, GM.

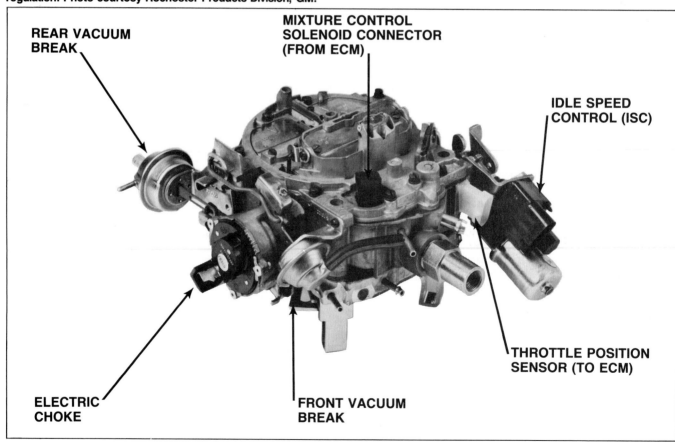

REAR VACUUM BREAK

MIXTURE CONTROL SOLENOID CONNECTOR (FROM ECM)

IDLE SPEED CONTROL (ISC)

THROTTLE POSITION SENSOR (TO ECM)

ELECTRIC CHOKE

FRONT VACUUM BREAK

ELECTRONIC CARBURETION

A new method of metering control was introduced in California passenger cars in 1980; in 1981 it went nationwide. The E4M series Q-jet, used with the GM Computer Command Control (CCC) System, has solenoid-controlled metering. In the mid-'80s these units were fitted to some light-duty trucks and commercial vehicles.

The E4M series differs from the original M4M carburetors in that the primaries have new features for optimum A/F mixture control during all ranges of engine operation. A *mixture control (M/C)* solenoid in the float bowl controls metering to the idle and main systems. Fuel metering is controlled by two stepped metering rods, operating in the primary jets, and positioned by a spring-loaded plunger in the M/C solenoid.

The plunger is activated (pulsed) by a signal from the *electronic control module* (ECM). It is a microprocessor controlling the CCC, which monitors engine parameters. The ECM receives inputs on coolant temperature, engine speed, carburetor throttle-blade position, intake-manifold pressure and oxygen content in the exhaust stream. This information is processed in milliseconds and the ECM sends outputs to engine components to control their function.

Two primary sensors supplying data to the ECM are the oxygen sensor and the *throttle position sensor* (TPS). The ECM responds to the oxygen sensor in the exhaust manifold by energizing the solenoid to depress (push down) the plunger—and rods—to lean the mixture, or de-energizes them—they move up—to richen it. The ECM supplies, on average, the ideal part-throttle A/F mixture ratios for efficient working of the catalytic converter. During heavy throttle the unit reverts to *open loop*—see following text—and more conventional metering control.

The TPS is mounted in the float bowl. It replaces the vacuum and mechanically operated throttle controls of earlier units. It supplies the ECM with throttle-position changes, which, in turn, adjusts the M/C solenoid accordingly.

Signal Processing—During warmup driving, the idle and main systems are controlled by signals from the ECM. During all driving ranges, it is programmed to pulse the M/C solenoid 10 times per second.

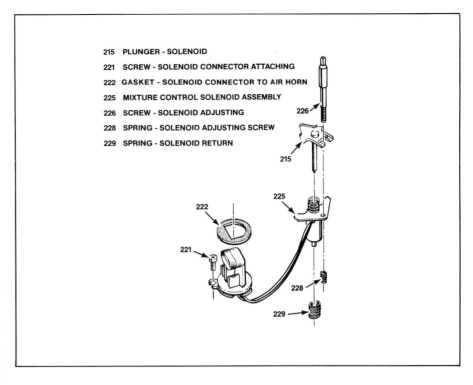

215 PLUNGER - SOLENOID
221 SCREW - SOLENOID CONNECTOR ATTACHING
222 GASKET - SOLENOID CONNECTOR TO AIR HORN
225 MIXTURE CONTROL SOLENOID ASSEMBLY
226 SCREW - SOLENOID ADJUSTING
228 SPRING - SOLENOID ADJUSTING SCREW
229 SPRING - SOLENOID RETURN

M/C solenoid (sometimes called the *dancing needle)* at top controls primary A/F metering under direction of ECM in *three- and four-point adjustment* carburetors. Some 1985/6 Q-jets (called *two-point adjustment* carburetors) have *rich limit stop* bracket (bottom drawing) controlling plunger travel upward to limit rich mixture. Drawings courtesy Rochester Products Division, GM.

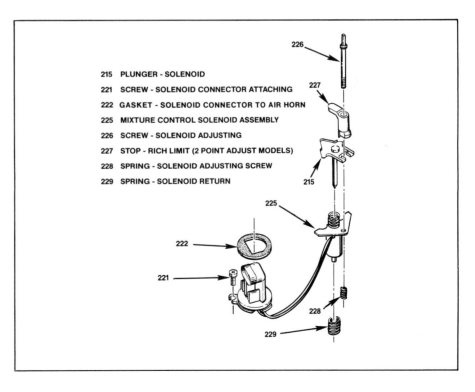

215 PLUNGER - SOLENOID
221 SCREW - SOLENOID CONNECTOR ATTACHING
222 GASKET - SOLENOID CONNECTOR TO AIR HORN
225 MIXTURE CONTROL SOLENOID ASSEMBLY
226 SCREW - SOLENOID ADJUSTING
227 STOP - RICH LIMIT (2 POINT ADJUST MODELS)
228 SPRING - SOLENOID ADJUSTING SCREW
229 SPRING - SOLENOID RETURN

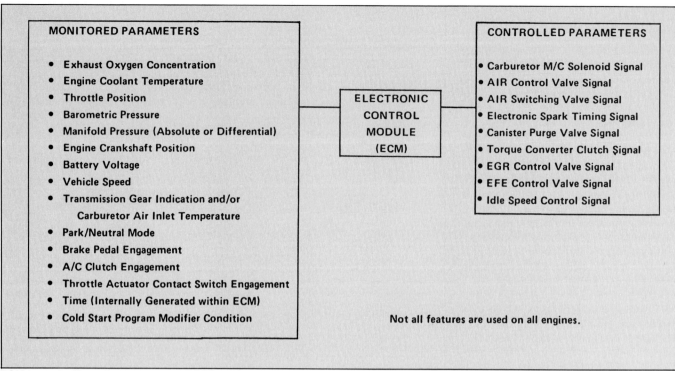

MONITORED PARAMETERS		CONTROLLED PARAMETERS

MONITORED PARAMETERS

- Exhaust Oxygen Concentration
- Engine Coolant Temperature
- Throttle Position
- Barometric Pressure
- Manifold Pressure (Absolute or Differential)
- Engine Crankshaft Position
- Battery Voltage
- Vehicle Speed
- Transmission Gear Indication and/or
  Carburetor Air Inlet Temperature
- Park/Neutral Mode
- Brake Pedal Engagement
- A/C Clutch Engagement
- Throttle Actuator Contact Switch Engagement
- Time (Internally Generated within ECM)
- Cold Start Program Modifier Condition

ELECTRONIC CONTROL MODULE (ECM)

CONTROLLED PARAMETERS

- Carburetor M/C Solenoid Signal
- AIR Control Valve Signal
- AIR Switching Valve Signal
- Electronic Spark Timing Signal
- Canister Purge Valve Signal
- Torque Converter Clutch Signal
- EGR Control Valve Signal
- EFE Control Valve Signal
- Idle Speed Control Signal

Not all features are used on all engines.

ECM is controlling microprocessor of CCC engine management system. Chart courtesy Rochester Products Division, GM.

TPS signals carburetor throttle position to ECM. Drawing courtesy Rochester Products Division, GM.

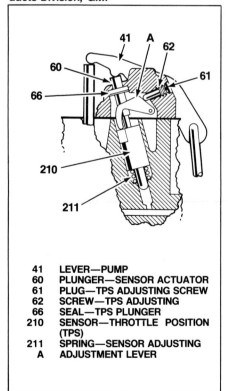

41	LEVER—PUMP
60	PLUNGER—SENSOR ACTUATOR
61	PLUG—TPS ADJUSTING SCREW
62	SCREW—TPS ADJUSTING
66	SEAL—TPS PLUNGER
210	SENSOR—THROTTLE POSITION (TPS)
211	SPRING—SENSOR ADJUSTING
A	ADJUSTMENT LEVER

METERING CONTROL SIGNALS

Engine Condition	Inputs to ECM	M/C Solenoid Operation
STARTING (Cranking)	Tachometer less than 200 rpm	M/C solenoid off (rich mixture)
WARM-UP	Tach above 200 rpm (engine running) O_2 sensor less than 315C (600F) Coolant less than 19C (66F) Less than 10 seconds elapsed since starting	Fixed command from ECM to M/C solenoid
WARM OPERATION, IDLE AND CRUISING (constant engine speed)	O_2 sensor above 315C (600F) Coolant above 19C (66F) MAP sensor	M/C solenoid signal determined by oxygen sensor information to ECM
ACCELERATION AND DECELERATION (changing engine speeds)	Throttle position sensor (TPS) MAP sensor O_2 sensor	Momentary programmed signal from ECM during period after throttle change, until oxygen sensor resumes control of M/C solenoid
WOT	TPS fully open MAP sensor	Very rich command to M/C solenoid

Locking the idle and part-throttle A/F mixture ratio at 14.7:1 could be more accurately accomplished if the M/C solenoid cycled faster. As designed, it can't adjust to this perfect ratio at 10 times per second, so it compromises.

When the ECM signals for more fuel, the plunger pulsates upward. Because it is moving up and down as it progresses to a richer position, it is actually sending an *average signal* of the required A/F mixture. So, the plunger doesn't instantly supply perfect mixture when directed by the ECM. It pulses up or down as it seeks richer or leaner averages by traveling upward or downward to meet the ECM's messages. The pulse rate is sufficient to meet today's emission and driveablility demands, so it's not likely to change.

Closed Loop Operation—A fundamental mixture-control signal to the ECM originates with the oxygen sensor. It supplies input—a voltage signal—to the ECM about oxygen content in the exhaust. The sensor compares oxygen content of exhaust against oxygen content of outside air— their difference creates voltage inputs to the ECM.

As oxygen content increases—leaning the mixture—voltage drops, the ECM de-energizes the M/C solenoid, and the metering rods move to the up (rich) position. When oxygen content decreases—a rich mixture—voltage rises and the ECM energizes the M/C solenoid to move the metering rods down to a leaner average position. The ECM regulates the *"closed loop"* between the oxygen sensor and the carburetor. Remember too, the ECM doesn't hold the mixture absolutely steady, it cycles the M/C solenoid plunger to an average correct mixture based on the signal from the oxygen sensor.

Air metering to the idle system is controlled by an *idle air bleed* valve in the air horn. The valve follows the M/C solenoid's plunger to control the amount of air bled into the idle system to lean or richen the idle and off-idle fuel mixtures. See page 95 for its adjusting data.

Open Loop Operation—The carburetor and entire CCC system are calibrated as close to 14.7:1 as possible for emission-control purposes, but during warmup and heavy throttle application, richer mixtures are required. These richer driving cycles are referred to as *"open loop"* because the oxygen sensor isn't supplying voltage to

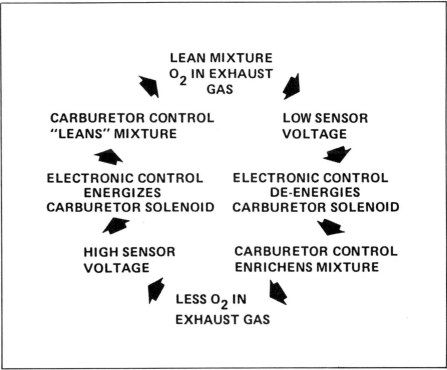

ECM determines proper fuel mixture by processing oxygen sensor voltage in closed loop operation. Drawing courtesy Rochester Products Division, GM.

Late-model (post-'81) Q-jet float system. Basic design is similar to earlier Q-jets. Drawing courtesy Rochester Products Division, GM.

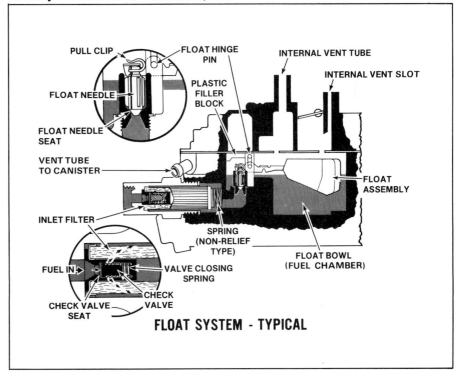

the ECM. Exact ECM metering is compromised during warmup and heavy throttle applications because the secondaries are not controlled to the optimum emission and fuel-economy A/F mixture ratios.

During warmup or until the coolant sensor reaches a certain temperature, ECM control of metering is based on engine requirements for a cold engine or heavy throttle operation. Its programmed instructions take over and control the metering for a richer mixture ratio. During conditions requiring richer mixture, the oxygen sensor doesn't participate.

It is important during heavy-demand driving that air-pump air is diverted from the catalytic converter by a diverter valve. The richer mixture will burn like a blowtorch in the heated catalyst when oxygen is pumped in with it. A glowing, red-hot catalyst can result. That can be dangerous when near combustible materials, and the excess heat will soon damage the catalyst. The ECM controls the diverter valve's activation. See page 171 in the Emissions Control chapter for more information.

NOTHING TO FEAR BUT FEAR ITSELF

Don't be afraid of electronic carburetion. Remember, the basic functions of a microprocessor-controlled carburetor are the same as those of a non-electronic one. Your initial fear may stem from the fact that electronic carbs are relatively new and have a reputation for being complicated.

Educate yourself at your own pace. Get a shop manual for a 1981 or later GM passenger car for reference and begin by reading its emission service section. You'll see that in reality there isn't that much new to learn about servicing these units.

There are three fundamental parts in an electronic carburetor you won't find on a conventional one.

Mixture Control (M/C) Solenoid—This controls the fuel-metering rods in the carburetor. By changing the duty cycle (electrical pulses) to the solenoid, the microprocessor establishes fuel mixture.

Throttle Position Sensor (TPS)—This is a sensor inside the carburetor, activated by the accelerator-pump lever. The TPS signals the microprocessor exactly how much the accelerator pedal is being pressed; that is, how far the throttle blades are open. The ECM adjusts the mixture according to throttle-position values stored in memory.

Idle Speed Control (ISC) Motor—On some models, this motor controls the engine-idle speed. It compensates for varying loads, such as the air-conditioner compressor and engine conditions. Not all cars have this control, but those that do require special care: You can't adjust the idle speed—the microprocessor alone regulates it.

TPS and main metering from electronic Q-jet: Note M/C solenoid doesn't have rich mixture stop to limit upward movement. Rich mixture is adjusted by screw. Drawing courtesy Rochester Products Division, GM.

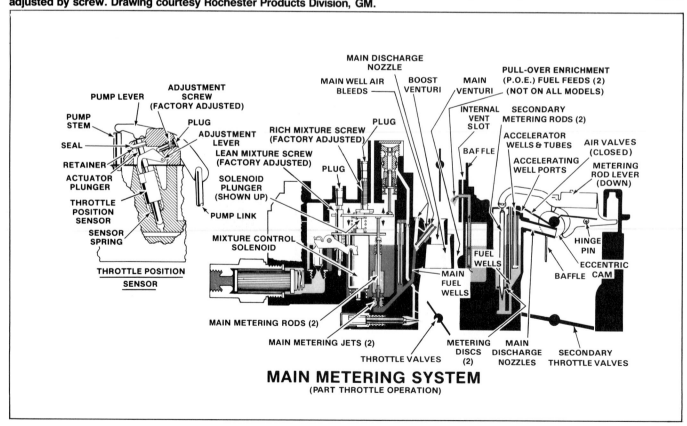

MAIN METERING SYSTEM
(PART THROTTLE OPERATION)

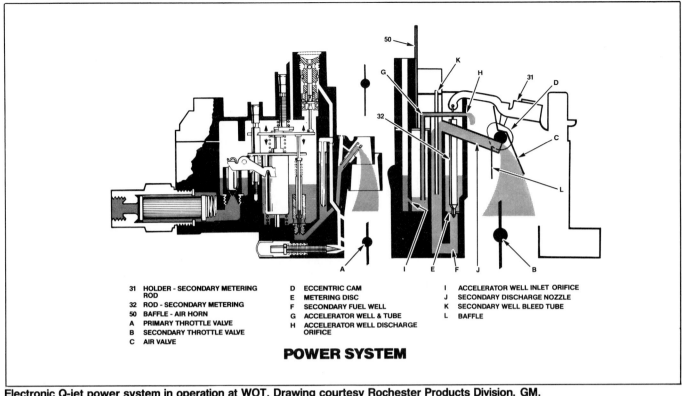

31	HOLDER - SECONDARY METERING ROD	D ECCENTRIC CAM	I ACCELERATOR WELL INLET ORIFICE
32	ROD - SECONDARY METERING	E METERING DISC	J SECONDARY DISCHARGE NOZZLE
50	BAFFLE - AIR HORN	F SECONDARY FUEL WELL	K SECONDARY WELL BLEED TUBE
A	PRIMARY THROTTLE VALVE	G ACCELERATOR WELL & TUBE	L BAFFLE
B	SECONDARY THROTTLE VALVE	H ACCELERATOR WELL DISCHARGE ORIFICE	
C	AIR VALVE		

POWER SYSTEM

Electronic Q-jet power system in operation at WOT. Drawing courtesy Rochester Products Division, GM.

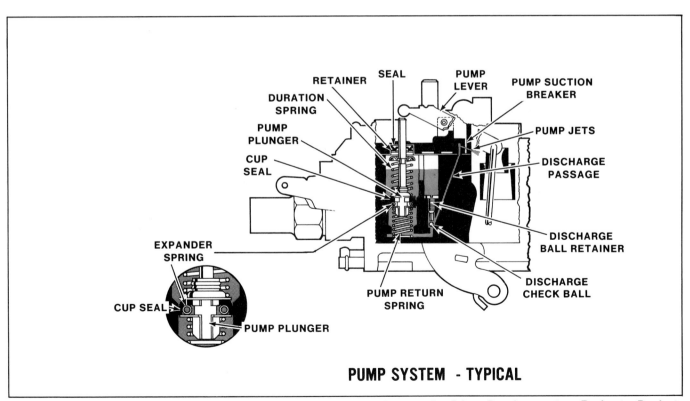

PUMP SYSTEM - TYPICAL

Accelerator pump system for most electronic Q-jets: Basic design is similar to earlier Q-jets. Drawing courtesy Rochester Products Division, GM.

PUMP SYSTEM - DUAL CAPACITY

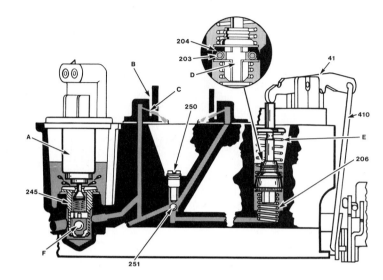

41 LEVER - PUMP
203 SPRING - PUMP PLUNGER
204 CUP - PUMP PLUNGER
206 SPRING - PUMP RETURN
245 VALVE ASSEMBLY - DUAL CAPACITY PUMP (SHOWN OPEN)
250 PLUG (RETAINER) - PUMP DISCHARGE

251 BALL - PUMP DISCHARGE
410 LINK - PUMP

A DUAL CAPACITY PUMP SOLENOID (SHOWN ENERGIZED)
B PUMP SUCTION BREAKER
C PUMP JET

D PUMP PLUNGER HEAD
E PUMP DURATION SPRING
F CHECK BALL

1985/86 Q-jets in U.S. trucks have dual capacity pump (top). When engine is cold, more fuel is needed for transition from idle to part-throttle. When engine is warm, less fuel is required. Dual capacity pump solenoid (bottom) is activated by coolant temperature sensor. At 170F (77C) pump solenoid energizes, dual capacity pump valve opens and pump capacity reduces by about one-half. Drawings courtesy Rochester Products Division, GM.

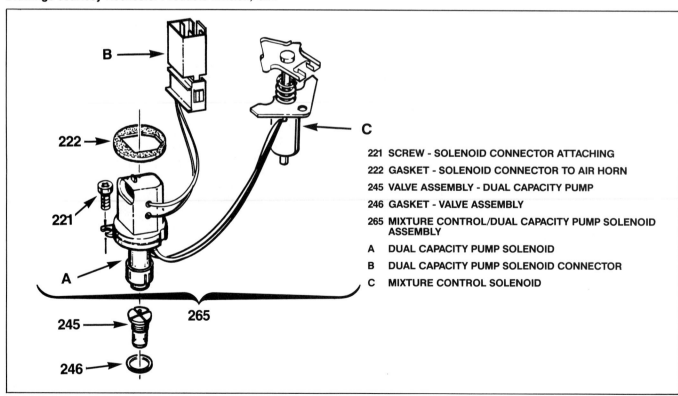

221 SCREW - SOLENOID CONNECTOR ATTACHING
222 GASKET - SOLENOID CONNECTOR TO AIR HORN
245 VALVE ASSEMBLY - DUAL CAPACITY PUMP
246 GASKET - VALVE ASSEMBLY
265 MIXTURE CONTROL/DUAL CAPACITY PUMP SOLENOID ASSEMBLY

A DUAL CAPACITY PUMP SOLENOID
B DUAL CAPACITY PUMP SOLENOID CONNECTOR
C MIXTURE CONTROL SOLENOID

Quadrajet Service

5

I believe the Q-jet has an unjust service reputation. All too often it is called "too complex, too complicated" by those who don't understand it. Not accepted as a high-performance carburetor by tuners, it also catches accusations by owners and mechanics who blame it for problems without diagnosing other potential troubles with equal fervor. Nevertheless, the Q-jet is superior to most carburetors for performance and general use.

I'm also convinced it's unlikely you'll ever have to replace an original-equipment carburetor. A carburetor is seldom the culprit in many tuneup problems. Also, you'll rarely come out ahead with a replacement unit instead of servicing your current one—if you do it correctly. To support this, remember the factory-installed carburetor has thousands of research and development hours behind it. Replacement manufacturers can't afford to spend a fraction of this time per vehicle or engine option. Assuming the proposed replacement unit is basically solid, the best you can get is a good carburetor calibrated to do a fair job. Check the list of engines it is recommended for. Do you really think it could be near-perfect for all of them?

Of course, carburetor service and overhaul are sometimes necessary, yet consider this costly task only after thoroughly troubleshooting simpler engine variables. I wrote this chapter to save you frustration and unnecessary expense. It is written and illustrated so you can master Q-jet service. If you have a fair knowledge of automotive mechanics, the explanations will seem unnecessarily detailed. These details will build confidence if this is your first Q-jet service job. Read the chapter, follow the illustrations and take your time to become a Q-jet expert.

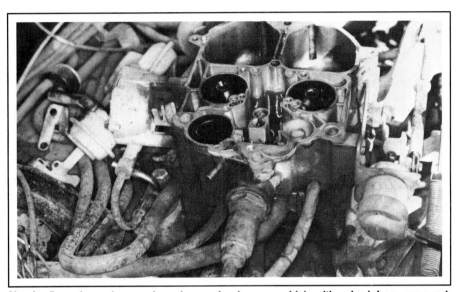

Nearly all repairs and general service can be done on vehicle with only air horn removed.

1967 Chevrolet Q-jet has air-deflector plate (arrow) beyond choke-blade housing. This was popular on several models. Plate directs airflow to stabilize bowl pressures during certain heavy-throttle maneuvers.

1972 Oldsmobile has entire choke assembly integrated with carburetor. Other Q-jet users mount choke thermostat (arrow) on manifold and connect it to choke blade by small rod.

1981 E4ME Q-jet with electric choke and microprocessor-controlled metering.

Model M4ME

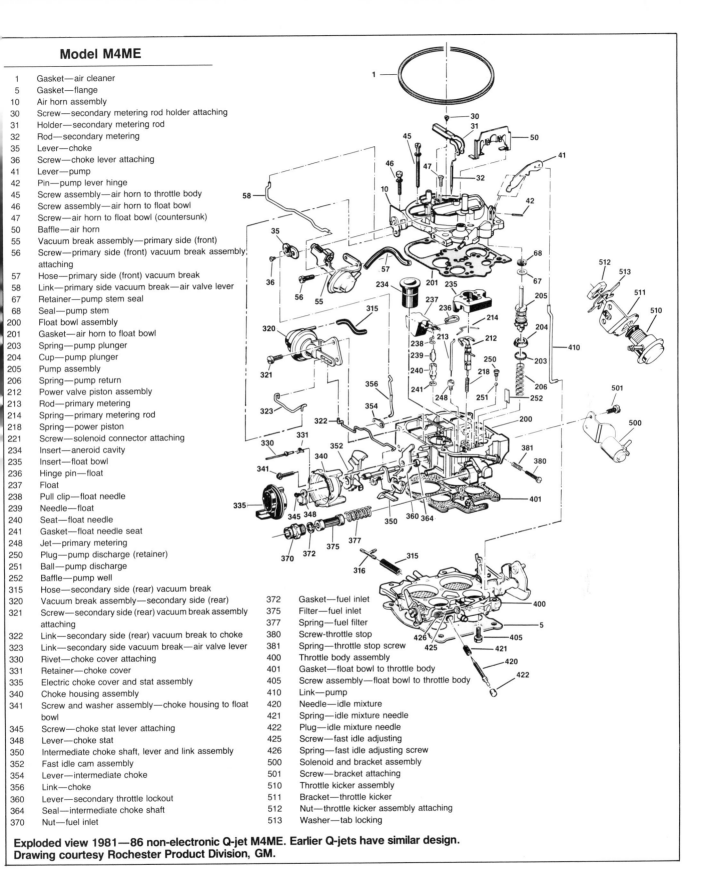

1	Gasket—air cleaner
5	Gasket—flange
10	Air horn assembly
30	Screw—secondary metering rod holder attaching
31	Holder—secondary metering rod
32	Rod—secondary metering
35	Lever—choke
36	Screw—choke lever attaching
41	Lever—pump
42	Pin—pump lever hinge
45	Screw assembly—air horn to throttle body
46	Screw assembly—air horn to float bowl
47	Screw—air horn to float bowl (countersunk)
50	Baffle—air horn
55	Vacuum break assembly—primary side (front)
56	Screw—primary side (front) vacuum break assembly attaching
57	Hose—primary side (front) vacuum break
58	Link—primary side vacuum break—air valve lever
67	Retainer—pump stem seal
68	Seal—pump stem
200	Float bowl assembly
201	Gasket—air horn to float bowl
203	Spring—pump plunger
204	Cup—pump plunger
205	Pump assembly
206	Spring—pump return
212	Power valve piston assembly
213	Rod—primary metering
214	Spring—primary metering rod
218	Spring—power piston
221	Screw—solenoid connector attaching
234	Insert—aneroid cavity
235	Insert—float bowl
236	Hinge pin—float
237	Float
238	Pull clip—float needle
239	Needle—float
240	Seat—float needle
241	Gasket—float needle seat
248	Jet—primary metering
250	Plug—pump discharge (retainer)
251	Ball—pump discharge
252	Baffle—pump well
315	Hose—secondary side (rear) vacuum break
320	Vacuum break assembly—secondary side (rear)
321	Screw—secondary side (rear) vacuum break assembly attaching
322	Link—secondary side (rear) vacuum break to choke
323	Link—secondary side vacuum break—air valve lever
330	Rivet—choke cover attaching
331	Retainer—choke cover
335	Electric choke cover and stat assembly
340	Choke housing assembly
341	Screw and washer assembly—choke housing to float bowl
345	Screw—choke stat lever attaching
348	Lever—choke stat
350	Intermediate choke shaft, lever and link assembly
352	Fast idle cam assembly
354	Lever—intermediate choke
356	Link—choke
360	Lever—secondary throttle lockout
364	Seal—intermediate choke shaft
370	Nut—fuel inlet
372	Gasket—fuel inlet
375	Filter—fuel inlet
377	Spring—fuel filter
380	Screw-throttle stop
381	Spring—throttle stop screw
400	Throttle body assembly
401	Gasket—float bowl to throttle body
405	Screw assembly—float bowl to throttle body
410	Link—pump
420	Needle—idle mixture
421	Spring—idle mixture needle
422	Plug—idle mixture needle
425	Screw—fast idle adjusting
426	Spring—fast idle adjusting screw
500	Solenoid and bracket assembly
501	Screw—bracket attaching
510	Throttle kicker assembly
511	Bracket—throttle kicker
512	Nut—throttle kicker assembly attaching
513	Washer—tab locking

Exploded view 1981—86 non-electronic Q-jet M4ME. Earlier Q-jets have similar design.
Drawing courtesy Rochester Product Division, GM.

REMOVING THE Q-JET

Cleaning—Begin with tidy working conditions by washing the engine at a car wash or with a suitable engine cleaner. Don't remove the air cleaner, and wrap the distributor with aluminum foil or plastic to keep water out. Don't forget to remove this waterproofing before driving.

Air-Filter Housing—Remove the vacuum line from the steel tube on carburetor. Leave the line attached to the air cleaner. Don't worry if the steel line pulls out of the carburetor. It's a press-fit and the hose usually sticks on the tube. Separate them and drive the tube back in with modest tapping. Place a screw in the tube to prevent damage as you tap the end.

Remove the wing nut and lift off the filter housing. The heat-stove pipe connecting the snorkle to the exhaust manifold is a slip fit. Ease it apart when lifing the housing. Some vehicles have a flex tube in place of the steel pipe. In some cases it's easiest to detach the bottom end of the pipe from the heat stove. Gentle turning and pulling will separate the two. Some heat pipes are clamped and must be loosened first. Most late-model vehicles use flex pipe—treat it with care and replace as needed.

Next, remove the vacuum hose located at upper front of the carburetor. If it does not pull off easily, I break the hose loose by prying with a screwdriver.

PCV Valve—Remove this large hose by spreading the squeeze clip with needle-nose pliers while gently pulling on the hose. Pry with a screwdriver if it doesn't come loose easily. Hoses get hard and brittle with heat and age; replace cracked or damaged ones. Note hose connections—both ends!—then use the hose as a sample when buying a replacement.

Throttle Cable—Remove the cable by prying off the spring socket that connects it to the throttle lever. This connection may vary from year to year, but removal will be obvious. On some cars the throttle linkage can't be pried off—a nut may hold the fitting to the throttle arm.

Transmission Kick-Down Cable—Remove cable by detaching the E-clip and slipping the elongated cable end from the throttle linkage. There will also be differences in transmission shift levers and cables among vehicles. Make a sketch of the hookup to aid reassembly.

Choke Rods—A screwdriver or needle-

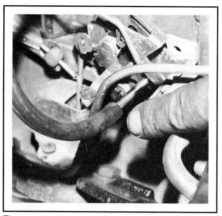

Remove vacuum line hoses from carburetor. Make sketch of all connections to aid reassembly.

Reinsert press-fitted steel line with modest tapping. Note screw in end prevents damage to line.

Lift off filter housing and separate hot-air connection between snorkle and exhaust manifold.

Vacuum lines sometimes resist removal. Break loose by prying with screwdriver or similar tool.

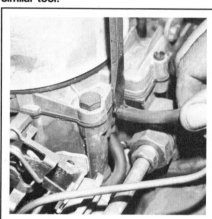

PCV valve hose requires pliers to remove squeeze clamp.

Remove choke rod retainer. Arrow indicates tang to be sprung outward off rod end so clip can be pulled free.

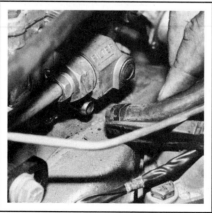

nose pliers will handle various retainers on choke-actuating rods. Once the rod is loose, let it stand free. There is no need to detach its lower end.

Fuel Line—Remove the fuel line with flare-nut wrenches. Because the fuel-line nut is large and often over-tightened, it may be difficult to loosen. If so, firmly hold the 1-in. carburetor fitting with an open-end wrench. Use a flare-nut wrench on the fuel-line nut and firmly tap it counterclockwise. Sharp blows will loosen a nut or fitting better than a strong-arm pull. Loosen the 1-in. fuel-inlet nut while the carburetor is on the engine.

Choke Mechanism—Early models secure the choke mechanism with a screw coated with thread-locking compound. Remove it now, leaving the carburetor on the engine. Use force with caution.

Base Capscrews—Remove the four base capscews with a 1/2-in. wrench. These secure the carburetor and various brackets—throttle-cable bracket for one—to the manifold. Detach these brackets as necessary to free the carburetor.

Throttle Return Spring—This hooks to the throttle-cable bracket; it doesn't have to be removed until the bracket is free. Release the spring from the throttle linkage, keeping it attached to the bracket and cable.

Electrical Connectors & Vacuum Lines—Remove as required. Take your time when detaching these to label all connections. You'll thank yourself when reassembling the carburetor.

Remove Carburetor—Lift off the carburetor. Heat, pressure and time often cement it to the manifold. Tap upward with a plastic hammer against a boss or other substantial part of the carburetor body to loosen it. Don't tap against any linkage, lever or bracket and don't pry against the carburetor base.

Now, inspect for air leaks around the base. It's not uncommon for the base gasket to deteriorate. A leak can cause idle problems, exhaust popping noises, whistles, and generally poor low-speed operation. This deterioration is accelerated by incorrectly torqued (loose) carburetor retaining nuts or capscrews.

Manifold Opening—Cover the opening immediately to keep out dirt. A clean shop cloth folded to size or cardboard cut to fit work fine. Secure the cover.

Fixture Mounting—Place the carburetor on a holding fixture for handling ease and

Fuel-line nut is often overtightened and requires firm taps to loosen. Use exact-fit flare-nut wrench on fuel line, or be prepared to round off nut.

Remove base capscrews with 1/2-in. box wrench.

Be sure screwdriver firmly grips screw holding choke mechanism; it can require effort to remove. It may require impact screwdriver.

to prevent damage to throttle plates (blades) during servicing.

AIR HORN REMOVAL

Idle-Stop Solenoid—Some units don't have this solenoid, but these two options work on those that do:

Option 1: With the attaching screw removed, the air horn can be lifted off with the solenoid's bracket secured to it. This procedure is preferable unless you plan to soak the primary components in cleaner. I recommend leaving it all intact and cleaning the air horn and attached pieces with a brush and mild cleaner.

Option 2: If planning to soak these pieces in cleaner, remove all solenoids, switches and non-metal parts. Whether you remove the necessary non-submersible items from their bracket or remove them with the bracket attached depends on your tools and ability. These items are usually held by a thin nut that may be secured with a locking device. In most cases removing the specific item from the bracket will be easier than removing the bracket.

Observe whether or not the switch or solenoid is adjustable and handle it accordingly. Measure everything and make notes for reassembly.

Manufacture carburetor stand for rigid mounting and suspension. You'll need one if doing plenty of tuning or rebuilding.

Arrows point out where air leaked around carburetor base gasket. Retaining nuts and capscrews were too loose.

Detach throttle return spring from throttle-cable bracket.

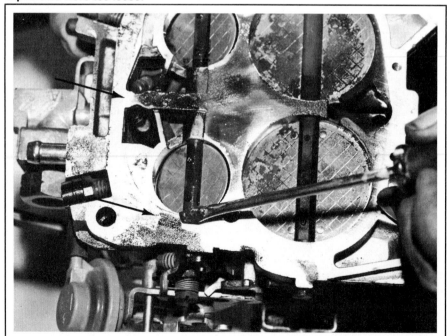

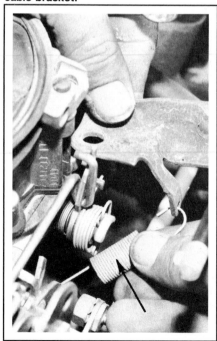

Four 5/16-in. bolts can be used to mount carburetor so throttle levers and blades are free to move.

First step in removing choke-blade linkage is getting this clip off rod.

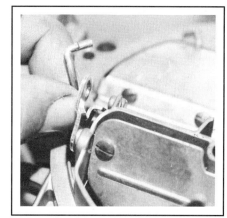

Next, pull rod free of choke lever.

Cutaway carb reveals lever you are trying to wiggle rod out of. Note rod scuff pattern (arrow).

Lift rod from cavity when it's detached from hidden choke lever.

Early Q-jets have simple pump-rod/lever attachment. Remove clip.

Internal Choke Blade Linkage—This step requires dexterity. First, remove the clip or retaining screw. Pull the choke rod free from the top lever. Sometimes prying is needed to do this. Next, free the rod from the choke lever hidden in the choke-lever well. The rod is not secured to the lower lever. Rotate the rod 90° and apply pressure to separate it from the lever. See cutaway view for how the rod fits into the lower link.

This link would normally be secured on a flat-sided shaft that connects to external choke components. There is minimal clearance for the rod to slip free of the link, so wiggle the rod as you pull up and turn it. This action encourages separation, but if it fails, insert a screwdriver in the well and gently force the lever outward. Lift the rod from the cavity once it is loose.

Note in the cutaway drawing how close the rod is to the inner wall—it has scuffed it. When secured at the top, the rod cannot turn and won't detach at the bottom.

Pump Rod—There are several types of pump-rod and pump-lever attachments. Note the type and follow the removal procedure accordingly. This is an important point in Q-jet servicing, so study this section and its photos before proceeding.

Some early Q-jets use a clip to attach the pump link to the pump arm. If your carburetor is of this type, you are in luck because the clip method saves time when disassembling and reassembling the air horn. Remove the clip and separate the rod from the pump arm.

On other Q-jets the pump arm is attached with its bent end protruding through the link. Disassembly is *not* simple; the pump

71

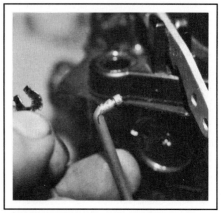
After removing clip, separate rod and lever. Easy, compared with later Q-jets.

The objective: drive out pivot pin and separate rod and lever.

File backstop keeps pivot pin from hitting air horn when driving out pin.

Use 1/32-in. drift to drive pivot pin through lever.

This screw attaches secondary-metering-rod hanger.

Secondary metering rods lift out with bracket.

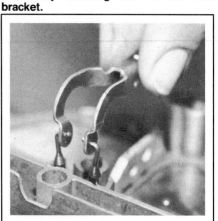

Gasket should stay on bowl as air horn is removed.

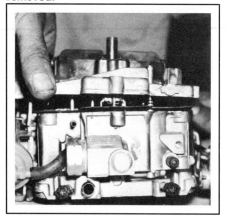

arm must be removed, and this requires some effort.

Brace a file or similar thin—0.04 in.—backstop between the vertical choke housing and pump-lever pin. Use a 1/32-in. drift punch to drive the pump-lever pin through the lever toward the choke. If the lever does not come free, the backstop is too thick—use a thinner one. If you inadvertently drive the pin tightly against the air-horn casting, use side-cutter pliers to push it back through the pump-lever hole for another attempt. Choose the backstop material carefully—too much force will cause damage.

Secondary Metering Rods—Remove the rods and their holding fixtures by unscrewing the small retaining screws. The rods will lift out with their holder. Remember their position—rod ends face to the carburetor center.

Air-Horn Screws—Remove all the screws holding the air horn to the bowl. Most screws are tightened firmly with a

Nine screws (arrows) attach early Q-jet air horn. Electronic carburetors have 13 screws.

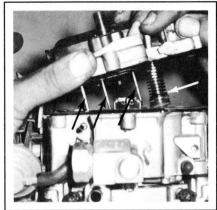

Two center tubes are idle tubes. Outer tubes (one hidden) are secondary feed tubes. Accelerator pump is noted by white arrow.

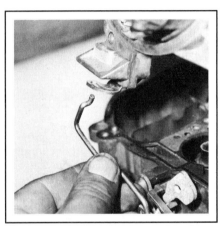

Tip air horn 90° to remove secondary damping link on mid-'60 to late-'70 models.

Arrow (A) indicates alignment pins. Lift gasket off these before trying to slide it from beneath metering rods at (B).

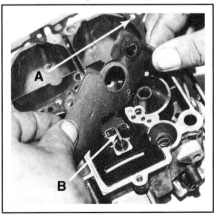

power tool during assembly. Many are secured with thread-locking compound, so use an appropriate screwdriver—it should fit tight and have a hard blade.

Two of the screws are hidden. Follow down along the choke blade to unscrew them. These two frustrate many mechanics who are unfamiliar with the Q-jet.

Don't be surprised if you find slotted, Phillips or clutch-head air-horn screws. The latter were introduced in the '80s to further discourage the do-it-yourselfer. Electronics stores, such as Radio Shack, and most tool suppliers will have the appropriate tools to remove these. Note that the head type and number of air-horn screws changed several times over the years. Other than the two hidden ones near the primary boost venturis, all are easy to extract.

Air Horn—Lift it straight off the bowl. Its gasket should stay on the bowl, but if it clings to the air horn, use a thin blade or your fingers to pry them apart. It's a good idea to have a replacement gasket available. Do not reuse a broken or obviously damaged gasket because well-sealed pas-

sages, channels and fuel compartments are critical to carburetor operation.

When lifting, keep the air horn centered until the protruding tubes clear their cavities. Careless handling can bend these non-serviceable items. On many models the horn must be tipped toward the choke mechanism. The accelerator-pump plunger may stay in its well instead of lifting out with the air horn, as illustrated.

Continue to lift the horn and tip it until it is perpendicular to the bowl. Now remove the secondary damping link from the air-valve lever. Early models with a fuel-damped air valve do not have this link. Set the air horn aside; turn it upside-down to prevent bending the projecting tubes. If it didn't lift out with the horn, lift the pump plunger from its well.

Air-Horn Gasket—First free it from the alignment pins and then from the rest of the carburetor body. Then carefully remove it from around the metering rods. It is easier if the gasket is first freed from the alignment pins at the rear. This allows it to slide under the metering rods.

If the gasket is in good condition—no cracks or tears—and you intend to reuse it, thoroughly lube it on both sides with heavy oil or light grease. This stops it from drying out; when reassembling, leave the lube on.

FLOAT BOWL DISASSEMBLY

NOTE: Don't lose any of these small parts—organize them in a tray. They may not be included in a rebuild kit.

Plastic Filler—Remove the plastic filler that covers the float assembly. Lift out the pump-return spring.

Power Piston—Next, remove the power piston and primary metering-rod assembly. To do this without tools, push the piston assembly down until it bottoms—then release it quickly. Spring force will push the piston up against its retainer with some striking force. Repeat this action five to 10 times and the assembly will free itself. *This is the preferred removal method,* unless a defective sticky piston damps the striking action.

If the piston is stuck, use needle-nose pliers to pry the assembly out of its well. Do not exert side forces. Attempts to pry the piston occasionally cause damage. If the metering-rod holding bracket separates from the piston, carefully remove the piston by using needle-nose pliers and *patience.* Join the two pieces back together after a thorough cleaning. A few drops of epoxy will assure a lasting repair. The power piston must be 1.950—1.955-in. long, not including the APT pin at the bottom of the piston. Measure and reassemble to this length. Clean the power piston with crocus cloth.

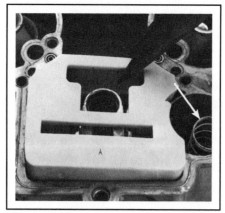
Lift plastic filler covering float assembly. Remove pump-return spring (arrow).

Push power piston assembly to bottom, then let your finger slide off it so spring shoves piston up against retainer. Do this five or ten times and assembly usually frees itself.

Try to pry stuck piston out of its well with needle-nose pliers.

If bracket separates from piston, epoxy can be used to make effective repair. Make sure hanger inserts into piston to its original depth as this affects metering-rod position.

Clean power piston with crocus cloth. Don't use sand or emery paper; they will scatch piston surface.

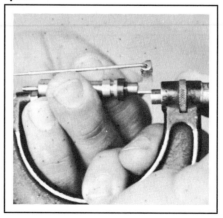
Measure 1.950—1.955 in. from hanger top to piston bottom (not including APT pin) to reassemble power piston that has pulled apart.

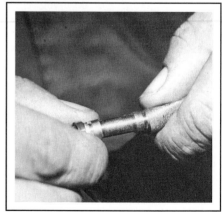

NON-SERVICEABLE ITEM
This term means RPD doesn't sell it as a separate part. They are also listed as NSS, meaning Not Serviced/Supplied Separately. If you damage a non-serviceable part, you must repair it, use an equivalent from a spare or junked carburetor, or buy a new assembly that includes the part. No Longer Available (NLA) is also a term for these parts.

Consequently, I've stressed cautious handling of the carburetor pieces as you disassemble it. Protect and organize pieces so they will not be damaged or lost as they await cleaning and reassembly. Care can save your money.

With your fingers or pair of pliers, lift float hinge pin straight up on 1967 and later models. Long hinge pin is retained in slots. Slots are different depths, so hinge pins must not be turned end for end. Observe it closely.

Hang needle pull clip on float as illustrated. On some floats it can be hung in pattern holes or from front. On others, this will create problems. Do it the safe way.

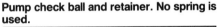

Pump check ball and retainer. No spring is used.

Some late models retained original short D-shaped hinge pin. Stamped *upsets* (arrows) in pin arch hold it from side movements.

Primary metering jets (A); screw (B) holds pump-discharge assembly.

Baffle in secondary side of bowl casting.

Float—On 1965—66 models with a diaphragm fuel-inlet assembly, lift the float up until the D-shaped hinge pin can be slid out to one side. Next, slide the float to the front until it slips clear of the small retaining clip, then carefully lift it out. If you're not careful, you may damage the clip or diaphragm assembly.

More disassembly is necessary with these models. With the float removed, loosen the two small screws and lift out the diaphragm assembly. If you intend to reuse the old diaphragm, handle it carefully and avoid using strong solvents such as carburetor cleaning fluid. Wash it off with kerosene or a similar safe cleaner when you are ready to reassemble it.

On 1967-and-later models with a needle-and-seat type fuel inlet, lift the D-shaped pin with your fingers or needle-nose pliers and the needle and float will follow. Note how the needle-to-float clip is mounted, and remember this for reassembly. To prevent needle-valve problems, the clip should stay on the needle when it is installed.

Remove the seat from the carburetor with the appropriate size screwdriver. The gasket under the seat is often difficult to remove without damaging it. Be careful if you don't have a replacement.

Primary Metering Jets—Remove these by putting a suitable screwdriver squarely in the jet and exerting a quick, counterclockwise turn. The brass jets will be damaged if the screwdriver slips. Next, extract the pump-discharge retainer screw. A check ball should be visible in the well. This is easily dumped out by inverting the carburetor. Don't lose it. They do come in some rebuilt kits.

Air-Direction Baffle—It is optional whether you remove this baffle from secondary side of the bowl. If it is tight, it's best to leave it. Its sides may vary in shape, so be sure it is installed as it was removed.

Secondary Metering Orifices—Do not remove these. They are stainless-steel washers permanently staked in place.

Throttle Body—Turn the carburetor upside down and remove the two—or three—Phillips screws holding the throttle body. Because these are steel screws in aluminum, they sometimes seize and require great force to break loose. An impact-driver may be the *only* way to get them out.

Lift the throttle body off the carburetor and remove the idle-mixture screws. On

Q-jet secondary metering orifices are not removable.

Cutaway secondary-jet area shows wafer-thin stainless-steel secondary metering orifice (arrow). Donut-shaped piece is crimped tightly in place above it as retainer.

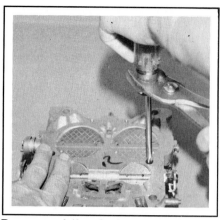

Recommended method of removing seized throttle-body screws: Note second person holding carburetor. Some of these require firm effort. Or, use impact driver.

Idle-mixture screw with spring and limiter cap.

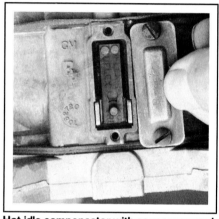

Hot-idle compensator with cover removed. It's usually located at rear of carb.

Bimetal valve and cork gasket of Q-jet hot-idle compensator.

When choke mechanism is removed this secondary lockout lever will be free to fall off (if carb has this type secondary lockout).

1971-and-later carburetors the idle-limiting caps must be removed before the screws can be backed out.

Hot-Idle Compensator—Remove the hot-idle compensator. Some Q-jets don't have them; they are usually at the back of the carburetor. Also remove the circular cork gasket in the well under the bimetal valve.

Choke—Remove the vacuum hose from the choke diaphragm. If the assembly is defective, remove the single screw holding it and slide it off. Watch for a secondary lockout lever that may fall off with the choke. Or, it could stay on its shaft, ready to fall off when you next tilt the carburetor. Turn the carburetor body upside down to recover the hidden choke link.

CLEANING

CAUTION: Any rubber, plastic, diaphragms and pump plungers should *not* be immersed in carburetor cleaner. The Delrin cam on the air-valve shaft can withstand *normal* cleaning in carburetor solvents.

Wear safety glasses and gloves when using cleaners—no matter how mild you think they are. Apply cleaners only in a well-ventilated workspace, keep them away from flame and avoid breathing the fumes.

If you choose to do a *complete* carburetor cleaning job for experience, thoroughness or cosmetics, put the submersible parts in a cleaning basket. After rinsing, clear all casting passages with compressed air. Do not pass drills through jets or passages.

Common early Q-jet choke with separate choke vacuum break and air-valve damper.

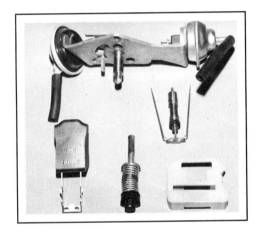

Don't put these parts or gaskets into carburetor cleaning solutions.

The throttle body, fuel bowl, air horn (shown) and small metal parts can go in this type cleaner basket.

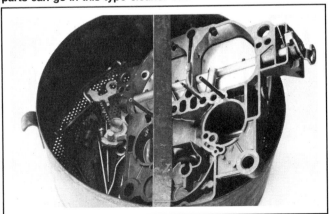

AVOID DISASSEMBLING CHOKE
Inspect it before removal, and understand it is the most complex mechanical linkage on the carburetor and can cause untold frustration. There are many variations in Q-jet chokes, and none are simple to take apart and reassemble. Cleaning with a spray cleaner/lube like WD 40 is all that is usually required.

Keep the work area clean and organized. Laying parts on white towels confines the small parts and makes all easy to spot.

The same suggestions for work-area cleanliness and safety apply even if you're doing a less complete overhaul. Cleaners such as kerosene, solvents and mild paint thinners can be applied to most delicate parts along with the metal pieces. A small open can, brush, small scraping tool and some elbow grease are fine for washing them.

Dissolve deposits with lacquer thinner or a similar cleaner. CAUTION: These must be used in a well-ventilated area away from fire. Don't breathe the fumes. Use cleaners with a brush on large surfaces to dissolve deposits, and use a syringe—a common ear syringe works fine—to shoot the solution through passages. Wear safety glasses and gloves when using these chemicals.

Bleeds & Vents—Only a few orifices and passages are subject to contaminant problems. Be sure the bleeds and vents are open in the air horn. You can see through most of them. If in doubt, run a small wire through. Orifices and bleeds in the air horn are not sensitive to minor changes that could be caused by cleaning with a piece of wire. The off-idle slots and idle-mixture needle orifice feed hole must be open and clean. Check with the needle screwed out.

Fuel Bowl—The fuel bowl contains critical metering orifices. Be careful about probing with wire or drills. Squirt liquid through the obvious fuel channels. These include jets, fuel-inlet needle orifice seat, pump channel and idle channels.

Power Piston Well—Be sure the power-piston well is clean. A twisted rag or a roll of crocus cloth will do a good job.

Throttle Blades—Move the throttle blades completely open and closed to be sure no foreign matter is binding the shafts.

Regardless of the cleaning method or materials used, wash the pieces thoroughly in hot detergent for final cleaning.

INSPECTION

Idle-Mixture Needles—Check for damage. Often, a needle has been turned into its seat too tightly, and damaged. The limiter caps on current carburetors prevent doing this. Pre-emission carburetors are most likely to have damaged idle-mixture needles and scored seats.

Float—Examine the float, needle and diaphragm for wear or damage. Be skeptical of the float's utility if it has over 40,000 miles or three years of service. They eventually absorb gas and become less buoyant. Replace if in doubt.

Power Piston—Be sure power piston is clean. Crocus cloth may be necessary if the piston is discolored or contaminated. Never use an abrasive that will scratch the piston's surface.

Carburetor Castings—Inspect upper and lower surfaces of carburetor castings for damage.

Levers—Inspect holes in levers for excessive wear or out-of-roundness. If con-

Some newspaper, old dishpan or open can, paint brush and some safe cleaner make suitable economy cleaning kit.

Ear syringe forces cleaners into critical passages to assure full openings.

Idle channels (A) and power-piston well (B) must be clean and free from deposits.

Use compressed air to blow through off-idle and curb-idle passages (arrows).

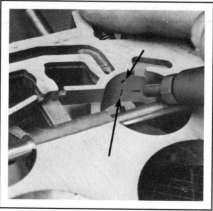

Power piston can be cleaned or polished with primary metering rods intact if care is used.

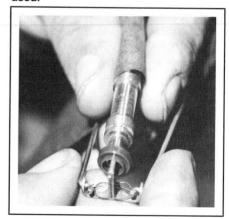

siderably worn, readjust corresponding links and rods. If wear compensation can't be made, replace or repair the lever. Replacement is rarely required. Check all throttle levers and valves for binding or other damage.

Fast-Idle cam—Examine fast-idle cam for wear or damage. Carefully inspect early plastic models for cracks. Replace if cracked or badly crazed from heat. Metal cams were fitted from 1967.

Air Valve—See if the air valve binds. If damaged, the air-horn assembly should be replaced. In most instances, it will take four to six weeks for your dealer to get these parts. A plastic cam and air-valve spring kit is available to fit all Q-jets—Part No. 7035344.

SPECIAL SERVICING
CHOKE

The Q-jet choke mechanism is relatively maintenance-free. Do not remove it unless a problem is evident—it is time-consuming to service. These parts deserve occasional observance and light maintenance.

External Linkage—The external linkage to the choke shaft can bind because of underhood contaminants. Vapors and dust form deposits that should be occasionally flushed and blown out of the pivot-bearing surfaces. The choke-shaft bearing surface on either side of the air horn can bind up with gum from evaporated fuel. A spray can or ear syringe with solvent will clean these quickly.

Vacuum-Break Diaphragm—These units normally give long service. Now and then a plastic housing is broken by carelessness, or badly crazed by underhood heat. Diaphragms can fatigue from age and use. A visual check tells their condition.

Beyond obvious wear and tear, check the break's function. Remove the vacuum hose from the housing. Push the lever into the diaphragm housing until it bottoms. Hold it firmly and close off the vacuum opening with your finger. Now, release the lever. It should move out a fraction of an inch—and hold its position while your finger seals the opening. If the lever moves, there is a small leak—replace the unit. If the diaphragm is good and holds, release your finger and let the lever fully extend. Some models have a purge bleed hole in the auxiliary vacuum-break tube. It must be covered or the diaphragm will not hold.

Check two dimensions when buying a replacement vacuum break. The lever must be the same size as the old one and the vacuum restriction should be similar. These restriction holes are sized accurately, so it's easy to measure them.

Put the point of a small sewing needle in the hole and mark the depth on the needle. Put it in the new unit to see how the depths compare. The orifice is generally 0.010—0.015 in. Occasionally one is 0.020 in. Be sure the old and new are *very close* in size.

This tiny restriction limits the speed the choke blade is pulled to its first position—vacuum-break setting—after a cold start. If it doesn't pull the choke blade partially open after cold-starting, extreme richness can result. Black smoke, engine stalling, poor gas mileage, and eventual damage to rings and cylinder walls from excess fuel diluting the cylinder lubricant can be caused by this.

The orifice has another important duty. On most models it regulates the air-valve opening speed. If not functioning correctly, secondary-throttle operation suffers. With no damping on the air valve, the engine will sag, hesitate and backfire under heavy throttle. Air-valve opening speed on 1965—1966 models is controlled by a lever-piston damping device.

Bimetal Choke Thermostat—Most of these mount on the intake manifold and are covered by a can. Many hours of design and testing determine the correct bimetal material, its length and number of coils.

Common problem with Q-jet chokes is binding shafts. Choke-blade shaft and choke-link shaft must be cleaned occasionally with gum-dissolving solvent.

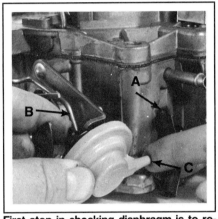

First step in checking diaphragm is to remove vacuum hose (A), push lever (B) into housing and hold your finger over vacuum opening (C).

If diaphragm is good, lever should move out only faction of inch and stay there so long as your finger seals vacuum opening.

Cover bleed hole (arrow) with tape before checking diaphragm on some models.

Check size of restriction if replacing vacuum-break unit. It should be very close to original dimension.

Always inspect power piston for deposits or marks (white arrows) indicating scuffing or binding in bore.

Cutaway showing key hooked to power piston metering rod crossbar (white arrow). See drawing page 81.

These establish the torque to close the choke blade and the rate it decreases with temperature.

Usual service is to be sure the coil housing is tight to the manifold and the can is in place. Some cans are held with a screw; others clip on with indented tabs. Be sure the rod is in place and not binding or rubbing the can or a hose. Hold the throttle partly open with one hand and move the choke blade from full open to full closed several times. It must be free.

Some Q-jets—typically Oldsmobile applications—have the choke thermostat in a housing attached to the carburetor main body. Be sure its parts are correctly attached and free from binding or wearing on other objects.

POWER PISTON

Inspection—Inspect for deposits or marks indicating scuffing or binding against its bore—a common problem with early-model, idle-vent-equipped Q-jets. In dry climates, the idle vent hole lets dust enter the piston well. Pre-'70 piston assemblies were held in the well by a metal spring clip, which didn't seal the piston from dust either. Most later models use a combination seal/retaining collar to hold the piston. It keeps out contaminants.

Problems—Power pistons occasionally stick because fuel contaminants and foreign material are forced into the piston area by engine-backfire pressures. Early Q-jets were plagued by this problem.

Backfiring can result from extremely lean part-throttle calibrations combined with chokes that disengage early to meet emission standards. The manifold and power-piston well are joined by a short vacuum passage. Some backfire flame front enters the piston's area through this passage. Usually, only fuel and air are blown in, but sometimes a troublesome power piston is found with exhaust contaminants on its surface: Carbon and flame are reaching it. The piston will eventually sieze.

If the piston sticks down, the primary power system won't function. The engine falters and hesitates at medium throttle. It will hesitate at warmup and for 2—3 miles after each cold start. If the piston sticks up, the mixture is rich for all driving conditions. Gas mileage will drop as much as 50%. Low-speed throttle response will be sluggish. Exhaust will be black and smoky.

Repair—Many a repair or carburetor replacement has been sold to fix this problem. A Q-jet troubleshooting pro can fix it in 30 minutes. A novice following instructions from this chapter can do it in less than two hours. Follow the accompanying photos on how to free a stuck power piston without removing the air horn.

You can make a special keylike tool by copying the one in the accompanying illustration. Insert it in the forward vent stack and hook the power piston/metering rod assembly crossbar. Gently push and pull the assembly—it should travel approximately 1/4 in. If it drags, spray in some solvent and work the piston up and down. Repeat this until its travel is free. Start the engine and run it briefly after each application of solvent to draw it along the assembly walls.

FUEL-WELL PLUGS

Plugs at the bottom of the secondary and primary fuel wells sometimes leak and cause rich mixtures at idle. This problem is cured or prevented by sealing them with epoxy. Delco and several parts companies sell special foam to fit under the secondary fuel wells as a fix. The foam is only partially effective and can cause throttle-body warpage if it is too hard. At best it only seals the secondary-well plug and leaves the small primary-well plugs to cause trouble. Epoxy them all and don't use the sealing pad.

THROTTLE BODY

Inspection—Hold the throttle body against a bright light and see if the throttle blades are correctly positioned in their bores. Do not adjust the blades unless they are *obviously* misaligned. Adjusting them is difficult to do correctly, so don't try unless you have an idle overspeed problem from air leakage, or a blade hangs up on the bore.

Adjustment—If you must—loosen the screws slightly and tap the blade with a small screwdriver or similar object as you exert closing pressure. Tighten the screws before releasing pressure. Some light around the blades is permissible; so if the blades look uniform in fit, leave them alone. RPD four-barrels seat throttle blades against throttle bores, not against an external stop—as is common on some other carburetors.

If piston happens to stick in up (rich) position, push it down with small screwdriver or use end of key. If that doesn't free it, extract with key.

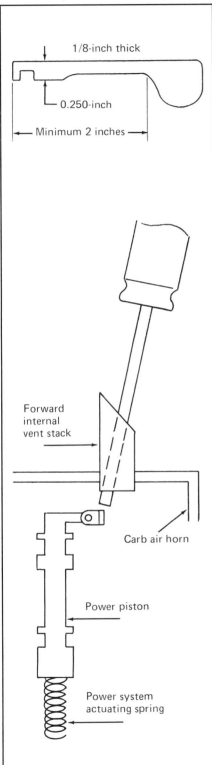

1/8-inch thick

0.250-inch

Minimum 2 inches

Forward internal vent stack

Carb air horn

Power piston

Power system actuating spring

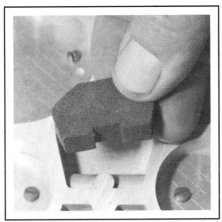

Foam pad designed to stop secondary fuel-well leakage is only partially effective at best. Epoxy is real answer here.

Secondary fuel-well (large) and primary fuel-well (small) plugs and surrounding area being cleaned prior to applying epoxy.

If you get this far into carburetor, use epoxy to seal plugs once and for all.

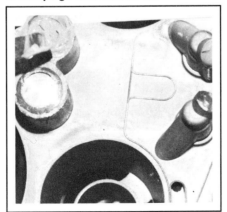

ASSEMBLY

Throttle-Body Gasket—Turn the bare carburetor fuel bowl upside down and fit the thick gasket over the alignment pins. An incorrect fit will be apparent because covered screw holes and interferences will prevent further assembly. Air leaks will result from an incorrect match, too.

Throttle Body—Place the throttle body on the bowl/gasket assembly and fasten the two or three screws. The bowl is light alloy so it is possible to strip the threads if the screws are over-torqued. If you are strong enough "to go bear hunting with a switch," use some restraint.

Idle Needles—Put the springs over the idle needles and screw them into the throttle body. Turn the needles in until they seat *lightly* and then back out two full turns.

Choke—Install the choke components next. Set the carburetor right side up on a fixture and proceed. Slide the fast-idle cam on the shaft of the choke linkage. The cam lobe and fast-idle steps will be toward the bowl. The fast-idle cam-adjustment tang should be *below* the cam.

If your carburetor has its secondary lockout link on the side of the bowl, set it on its pivot boss and engage the lockout pin on the secondary-throttle shaft.

Slide the choke linkage assembly with fast-idle cam along the metal vacuum tube until the end of the shaft is barely inside the hole in the side of the carburetor. To let the fast-idle cam pass the secondary lockout, pivot the lockout link counterclockwise for clearance.

Early model brass plugs sealed with epoxy. Not pretty, but effective.

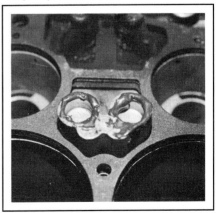

Hang the intermediate choke lever from the choke-actuating rod and lower into the passage on the side of the carburetor. The link should be suspended in the cavity until you can see it centered in the window hole.

Push the linkage assembly into the carburetor so the end of the large flat-sided shaft tries to enter the intermediate dangling lever. Pin the lever against the inside wall of the passage by pushing in with assembly. Use the choke-actuating rod to slowly rotate the lever counterclockwise until the shaft lines up with the lever hole and mates. Next, slide the linkage until it is flush with its mounting points on the carburetor and fasten with the single screw.

Temporarily remove the choke-actuating rod from the intermediate lever by rotating it counterclockwise and lifting it up out of the cavity. Removing it will make it easier to install the air horn.

If throttle blades are badly misaligned, they won't close correctly and large amounts of light will be visible around them. These are fine.

Incorrect placement of throttle-body-to-bowl gasket will cover screw holes and leave large manifold vacuum leaks. Place gasket on fuel bowl first.

Place throttle body on bowl/gasket assembly.

Turn screws snugly to seat gasket, but don't overtighten.

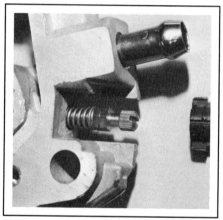

Needle installed in throttle body with spring in correct position. Idle-limit caps are no longer used for service. If you cut caps off, leave them off.

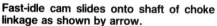

Fast-idle cam slides onto shaft of choke linkage as shown by arrow.

Where secondary lockout is on side of bowl, slide lever onto pivot so lockout pin on secondary throttle shaft is engaged (arrow).

Choke linkage and fast-idle cam being started over metal vacuum tube.

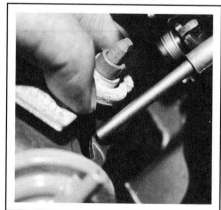

Vacuum Break—Connect the vacuum-source tube to the vacuum break. Engage the hole in the vacuum-break rod with the end of the air-valve damper rod. The upper end can be laid over the choke linkage until the air horn is installed.

Hot-Idle Compensator (if used)—Place the round cork gasket in its well in the hot-idle compensator chamber.

Bimetal Valve (if used)—Slide in the bimetal valve and fasten the cover down with two screws. This only applies to carburetors with a hot-idle compensator.

Accelerator-Pump Check Ball—Drop it into its well and screw in the retaining screw. There is no gasket under the screw.

Primary Jets—Screw the two primary metering jets into the bottom of the float bowl.

Accelerator-Pump & Power-Valve Springs—Place in their respective wells.

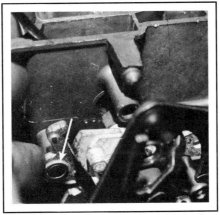

Secondary-lockout link pivoted to allow fast-idle cam to slide past.

Lower intermediate choke lever into passage until flat-sided hole is visible through side hole in carburetor main body (arrow).

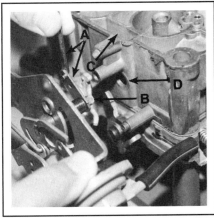

For illustrative purposes, intermediate choke rod and lever (A) are positioned on shaft (B). When actual assembly is done, link and lever will be dangled down in cavity (C) and shaft will enter hole (D) and join them.

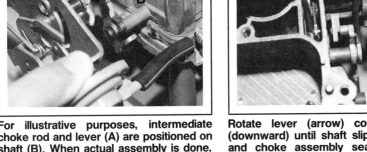

Rotate lever (arrow) counterclockwise (downward) until shaft slips through slot and choke assembly seats against its mounting pads.

Engage hole in vacuum-break rod with air-valve-damper rod.

Typical complete choke assembly.

New cork gasket in place in HIC.

HIC bimetal in place in cavity.

Fuel-Inlet Seat—Screw the fuel-inlet valve seat—with gasket—into the float-bowl assembly. If the fuel-inlet valve is one of the early-model diaphragm units, drop it into its well—position it carefully—and fasten it with two screws.

Needle—Drop the needle into the seat on late-model carburetors. Be sure the float pull clip is still attached to the needle.

Float—Hook the float bracket with the needle clip. Insert the float-pivot pin through the two holes in the float and drop the assembly into its slot. There are two float-pivot-pin versions. The short one (3/4 in.) can point in either direction when mounting the float. Install the longer (1-1/4 in.) pin with the open end toward the accelerator-pump well. In each case, the plastic stuffer will usually dictate correct pivot-pin orientation.

Float Level—Hold the rear of the float down on the fuel-inlet valve and check the float level with the gage furnished in the rebuild kit. If adjustment is required, bend the float arms at the adjustment points next to the power-valve well. Use rebuild-kit or shop-manual specifications for the correct dimension.

Primary Metering Rods—NOTE: the plastic stuffer that covers the float-pivot assembly has to be inserted *before* the metering rods on many Q-jets from the mid-'70s to mid-'80s. Carefully guide the two rods into the jets while lowering the power valve into its well. *Don't force the metering rods because the thin tips bend easily.*

Power Valve—On late models, use a wide screwdriver blade to force the plastic lock/seal into the top of the power-valve well. Some models have no top seal. If the power piston requires more than light finger pressure to push it down, check the spring to make sure it has guided up into the power valve correctly. Too much spring pressure will not allow manifold vacuum to pull the metering rods down in the jets. Poor fuel economy results.

Plastic Stuffer—See NOTE at Primary Metering Rods. Drop the plastic stuffer into position. The top of the stuffer shouldn't be higher than the top of the float bowl. If there is interference, the float-pivot pin may be installed backwards.

Air-Horn Gasket—Compare the new air-horn gasket to the top of the float-bowl assembly or old gasket and install it. A small flap goes under the T on the power

Partially reassembled float bowl assembly showing: (A) check ball retaining screw, (B) primary jets, (C) fuel-inlet valve seat, (D) accelerator-pump spring and (E) power-valve spring.

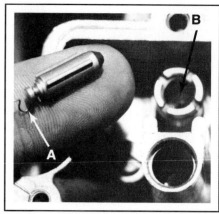

Needle with float pull clip (A) in place. Place in fuel-inlet seat (B).

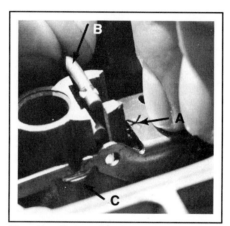

Float installation showing: (A) inlet-needle pull clip on power valve (inboard) side of float housing, (B) first-design (short) pivot pin and (C) float level adjustment bending point.

Be careful when inserting primary metering rods and power valve. Don't bend rod tips.

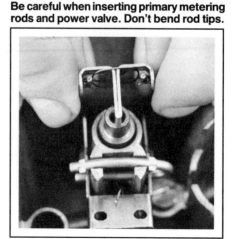

Set float level using gage supplied in rebuild kit.

Install plastic seal/lock/retainer for power-piston assembly into top of power-valve well with flat screwdriver blade.

Installed power-valve assembly: (A) plastic lock (retainer) and (B) metering rods. Be sure small tension spring is in place between metering rods.

Proper method for inserting air-horn gasket under primary metering rods. Note top of plastic stuffer is flush with top of float bowl.

Wrong way to place gasket on bowl assembly. Power valve will never work with gasket like this.

valve. Slide this flap under it and position the gasket. The power valve won't work unless the flap is placed correctly.

Accelerator Pump—Insert the accelerator-pump rod through its hole in the air horn. If not replacing plastic skirt or pump assembly, spread out the skirt with your thumbs. This assures a positive seal against the pump-cylinder wall. As long as the pump cup is not brittle, it can be rejuvenated for further service. If shrunk or brittle, replace.

Air Horn—Hold the end of the accelerator pump and tilt the air horn 90° to engage the air-valve damper rod. Slowly rotate the air horn down to the carburetor and make sure the four brass tubes fit their respective

holes and the accelerator pump fits into its well.

Insert its screws—9 to 13, depending on model year—and fasten the air horn to the float-bowl assembly. Don't forget the two hidden screws near the primary nozzles. Start all screws by hand before tightening from the horn center out.

Accelerator-Pump Linkage—Fit the bottom of the accelerator-pump rod into its hole in the throttle linkage. For carburetors with a single right-angle bend at the top of the pump rod, rotate the rod into the hole in the pump lever for your application. Push the C clip over the tip of the pump rod and crimp with pliers.

If the pump rod has an S bend at the top,

you have removed the pump lever during disassembly. Select the pump-rod hole you want and put the lever on the rod. Align the small hole in the lever with the pivot pin and press the pin through with a screwdriver. It requires dexterity to get the pin to start through the lever-pivot hole. Sighting through the hole sometimes helps.

Accelerator Pump Action—There are two pump-rod holes in some pump levers which establish pump travel. One at the end of the lever is used for normal driving in moderate climates. The inside hole, standard location on some models, gives a larger pump shot for racing or cold weather.

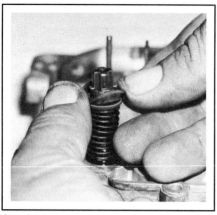

Accelerator pump correctly positioned on air horn. One method of "reconditioning" old pump skirt is shown. If there are cracks in plunger, replace it.

Accelerator-pump rod has upset tang that mates with keyed hole in throttle lever.

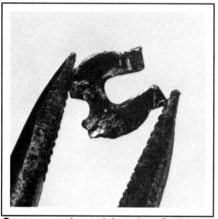

On some early models, crimp C clip onto accelerator-pump rod after engaging rod into desired hole in pump lever.

On other models, S bend at top of pump rod means you have to locate pump lever so pivot pin can be pushed through it.

Use screwdriver to push pivot pin through pump lever.

Getting choke rod installed is not always easy. Some force may be needed.

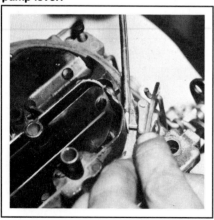

Choke—Insert the choke rod in the side cavity and connect it to the intermediate hidden lever. Rotate the outside choke lever clockwise and hold it firmly. Grip the top end of the rod and hook it in the lever. Look down the cavity to see how to turn and guide the rod to accomplish this. When hooked, release the outside lever and move the rod up and down—the outside linkage should move. This exercise can be frustrating. Wedging the hidden lever out against the cavity with a screwdriver can sometimes help.

Connect the other end of the rod to the choke-blade lever. Some force may be necessary to slip the rod through the hole. Fasten with its clip.

Secondary Metering Rods—Hang the rods from their hanger and drop them into the carburetor. Fasten with a single screw through the top of the hanger. See photo for correct installation in hanger.

Solenoid—On late model carburetors with a CEC (Combination Emission Control) solenoid, slide the solenoid bracket over the key on the air horn and rotate it down over the locking block. Bend the tangs around the bottom of the block to complete the installation.

INSTALLING

Carburetor to Manifold—Uncover the intake manifold, remove the old gasket and clean the mating surface. Put the new gasket on the manifold, using the two studs as a guide. Set carburetor on the gasket and put in the front mounting bolts. Turn these two bolts until they are slightly more than fingertight. NOTE: Early models with heat crossover use a stainless-steel shim next to the throttle body. A gasket fits between the shim and the manifold. The nut on the right rear of the carburetor should also be turned slightly more than fingertight.

Hook the throttle-return spring through the eye in the cable-supported bracket. Slip the bracket over the left-rear stud and tighten the nut to 10—14 ft-lb. Tighten the three mounting bolts to the same torque.

Each installation will have a slightly different bracket mounting, so install mounting capscrews and nuts accordingly—don't overtighten them.

Automatic Transmission & Throttle Cables—Attachment methods for these two cables vary from vehicle to vehicle. If the car has an automatic transmission, connect the detent cable with its clip. Snap the

Choke rod is secured in lever with E clip like this one, or with C clip.

Install secondary metering rods with rods inserted into hanger as shown here.

Rotate CEC (Combination Emission Control) solenoid bracket into place on keyed boss and fasten with single screw.

Installing throttle-return spring. Note that spring is attached to bracket before securing bracket to stud with nut.

Installing E clip onto automatic-transmission linkage. Attachment methods vary on vehicles.

Socket-and-ball assembled correctly on throttle cable. Pointed out here by screwdriver.

socket at the end of the throttle cable onto the ball on the throttle lever, then secure it with its clip.

Vacuum Hoses—Attach the PCV vacuum hose. Connect other vacuum lines per sketch made during disassembly. Vacuum connections change from model to model, so there is no absolute guide to what routes where. One hose often missed is the one to the temperature-control valve in the air cleaner.

Thermostatic Choke Coil—Connect it to the choke linkage.

Solenoid—Plug the two vacuum hoses and the wire plug into the CEC solenoid. No vacuum hoses attach to the CEC on manuals. The solenoid governs vacuum advance and deceleration control.

Fuel Line—Use a flare-nut wrench to tighten the fuel-line fitting. Tighten the fuel line modestly, set the air cleaner on the carburetor as a flame arrester in case of backfire, and start the engine. Seconds after it starts and runs, shut it off. Examine for leaks at the fitting nut; if there is seepage, tighten a little more and repeat the check.

Air Cleaner—Mount it and attach all vacuum lines and intake hoses.

SETTING IDLE CAN BE DANGEROUS

Use caution when setting idle. Many cars have automatic transmissions and final idle settings are made in DRIVE gear. If you are leaning under the hood and open the throttle quickly to note engine response, the car may leave with you draped over the fender or grill. Do not take this warning lightly, as some experienced mechanics have been pinned against walls and workbenches with badly crushed legs. Emergency brakes will not hold a heavy-throttled engine.

Play it safe—tune the engine enough in NEUTRAL to drive to a shop for final adjustments. If you insist on completing the job, drive the car's front bumper up against a tree or similar stout object and proceed with adjusting. *Don't rely on chocks or blocks under the tires to hold the car.*

Reattach PCV hose to fitting on Q-jet base.

This hose controls temperature valve in air cleaner. Don't forget to attach it.

Reattach two vacuum hoses and wire plug on CEC to put in service.

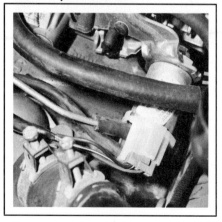

PRELIMINARY ADJUSTMENTS

If you need complete adjustment procedures for a particular carburetor, refer to the shop manual for that car or truck, or get them from your Delco dealer.

IDLE

For 1971-and-later models, set idle according to the tuneup sticker under the hood. Earlier models should be set by turning the idle screws clockwise in approximately 1/8-turn increments. Alternate from side to side until a lean roll—modest roughness—is noted. Back them out about 1/8-turn from this point. Be sure the idle speed is close to correct when final mixture setting is attained.

HIC VALVE CLOSING

Making idle adjustments requires closing off air entry if the HIC valve is open. Plug it or push its pin, depending on type, to do this.

FAST IDLE

Fast-idle rpm in neutral with a warm engine should be 1600—2400, depending on make and model. No matter what it is *supposed* to be, one specification cannot be correct for all drivers and geographical locations. Decrease or increase it to suit yourself. Don't judge it hot—it always sounds fast—that is, 2000 rpm when the engine is hot will be about 1000 rpm when it is cold. Make a small change, then evaluate it.

With the engine warm and off, push the throttle partly open, close the choke blade completely, and hold it closed as you release the throttle. Start the engine in neutral—with the brake on and vehicle immobilized—without touching the throttle, then check rpm on a tachometer. This is high-step rpm. Tap the throttle lightly to get if off the high step.

This adjustment is critical to engine operation immediately after a cold start. If the engine starts well but runs too slow while still on the high step, increase the warm-engine setting in 200-rpm increments until you are satisfied. If the engine "revs to the sky" right after a cold start, decrease the warm-engine setting in similar increments.

PUMP ROD

In many cases, whether or not adjustments are set to specifications from a manual is unimportant. Simply tailor the pump to fit your needs. Accelerator-pump requirements will be greater in a cold climate. To get ultimate performance and drivability it may be necessary to increase pump capacity in winter and reduce it in summer.

Make the following observation with the engine off. Look in the primary venturi with a flashlight. A stream of fuel from the pump shooter should flow each time the throttle is opened. If fuel is shooting, then it's a matter of increasing or decreasing capacity.

Moving the rod from the outer hole to the inner hole increases capacity. Bending up the lever tab also increases it; bending it down decreases capacity. For street use, don't adjust the bendable tab on the end of the lever in more than 1/32-in. increments.

IDLE-VENT VALVE

Vent valves used on pre-'69 models allow vapor to exit the fuel bowl during hot soaks and let in fine dust. If the vehicle restarts well after short hot soaks, it's either working or not needed. Bend the tab and close existing valves when possible.

If hot restarts are difficult, set the vent valve off the seat 0.020—0.040 in. when the primary throttle is at curb idle. A paper clip has a thickness of about 0.035 in.

VACUUM BREAK

The vacuum-break setting influences engine operation immediately after starting and for the first minutes of driving. If the car spits, coughs, sags and wheezes, bend the vacuum-break rod or tang for less choke-blade opening at full diaphragm travel. Increase the blade opening if the car loads up (runs rich), look for black smoke from the exhaust and slow restarts when the engine stalls.

This setting also is influenced with subtle adjustments. Work in 0.015—0.025-in. increments.

AIR-VALVE DASHPOT

The gap between the link and end of air-valve-lever slot is only important if sagging at high speeds occurs. Check a shop manual for adjustments for specific vehicles.

CHOKE UNLOADER

This helps overcome hot-starting problems. Holding the accelerator on the floor opens the choke blade a prescribed amount.

The open throttle also lets in air and the chamber is soon purged of extra fuel. Make the prescribed WOT unloader adjustment and use it during hot restarts. Remember to hold the throttle open during starting. *Never* pump the accelerator when starting a warmed-up engine.

AIR-VALVE & SECONDARY THROTTLE-VALVE LOCKOUT

Be sure this lockout arrangement is adjusted as specified, or secondary operation may be hampered.

CHOKE-COIL ROD

Cool the choke coil to observe the direction the coil pushes the blade closed. Study the carburetor to see to lengthen or shorten the rod for increased or decreased choke.

SERVICING ELECTRONIC Q-JETS— E4ME/C

Much of what has already been explained in this chapter applies to the non-electronic units used before 1981. The major design difference in post-'81 Q-jets is microprocessor-controlled fuel metering in the primary side. The following comments detail simple service procedures.

Of course, there is the option of getting a GM service or shop manual, buying an expensive rebuild kit and bucket of carburetor cleaner, and doing a complete teardown. That extensive job is seldom required. The new Q-jets are durable and can give years of service with minimum expense. Most problems with them, regardless of year of manufacture, have been caused by bungling mechanics.

Ordinarily, I don't need an overhaul kit for more than one out of every fifty units I repair. Perhaps one in twenty will need an air-horn gasket because the capscrews were tightened too much or the gasket was mishandled during a previous job. Every three or four years of service, a float may need replacing. A Throttle Position Sensor (TPS) or other electronic component may fail; but that, too, is relatively rare.

As emphasized before, it is usually unnecessary to remove the Q-jet from the vehicle and seldom necessary to remove more than the air horn for many service operations.

Nearly all important electronic carburetor (Q-jet or Dualjet) service can be done on vehicle with just air horn removed.

Part most often damaged (bent) during service is TPS stem, noted by pencil tip. Use no tools, handle with care.

Exploded view 1985/86 Q-jet E4MED used in U.S. trucks: Note dual pump capacity solenoid (#265) and related parts. Drawing courtesy Rochester Products Division, GM.

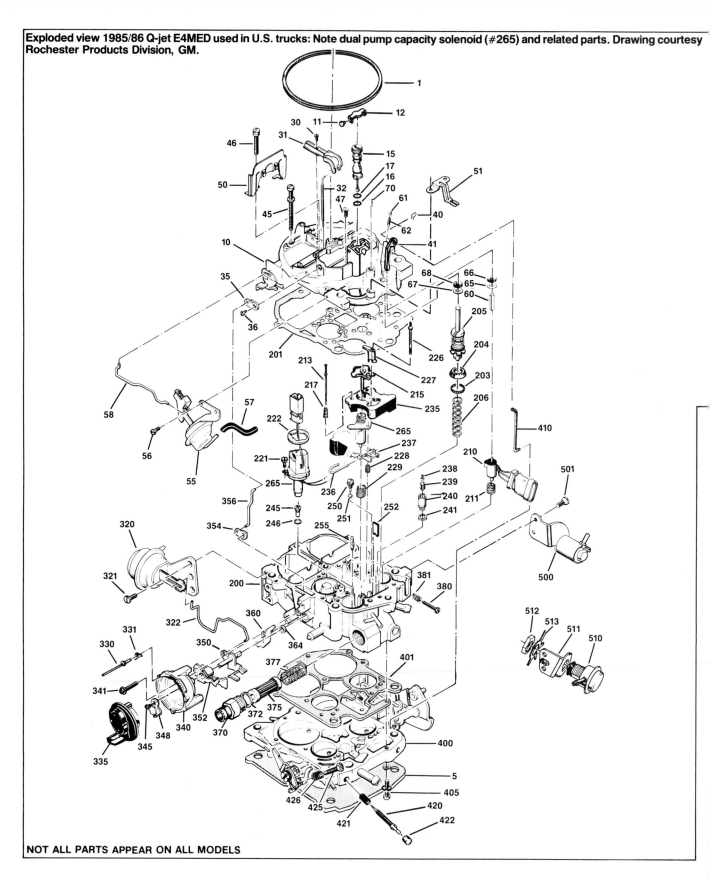

NOT ALL PARTS APPEAR ON ALL MODELS

Model E4MED

1	Gasket—air cleaner
5	Gasket—flange
10	Air horn assembly
11	Rivet—cover attaching
12	Cover—air bleed valve
15	Air bleed valve assembly
16	O-ring—air bleed valve—lower
17	O-ring—air bleed valve—upper
30	Screw—secondary metering rod holder attaching
31	Holder—secondary metering rod
32	Rod—secondary metering
35	Lever—choke
36	Screw—choke lever attaching
40	Retainer—pump link
41	Lever—pump
45	Screw assembly—air horn to throttle body
46	Screw assembly—air horn to float bowl
47	Screw—air horn to float bowl (countersunk)
50	Baffle—air horn
51	Retainer—mixture control solenoid assembly connector
55	Vacuum break assembly—primary side (front)
56	Screw—primary side (front) vacuum break assembly attaching
57	Hose—primary side (front) vacuum break
58	Link—primary side vacuum break—air valve lever
60	Plunger—sensor actuator
61	Plug—TPS adjusting screw
62	Screw—TPS adjusting
65	Retainer—TPS seal
66	Seal—TPS plunger
67	Retainer—pump stem seal
68	Seal—pump stem
70	Plug—solenoid adjusting screw

200	Float bowl assembly
201	Gasket—air horn to float bowl
203	Spring—pump plunger
204	Cup—pump plunger
205	Pump assembly
206	Spring—pump return
210	Sensor—throttle position (TPS)
211	Spring—sensor adjusting
213	Rod—primary metering
215	Plunger—solenoid
217	Spring—primary metering rod (E2M, E4M only)
221	Screw—solenoid connector attaching
222	Gasket—solenoid connector to air horn
226	Screw—solenoid adjusting (lean mixture)
227	Stop—rich limit
228	Spring—solenoid adjusting screw
229	Spring—solenoid return
235	Insert—float bowl
236	Hinge pin—float
237	Float
238	Pull clip—float needle
239	Needle—float
240	Seat—float needle
241	Gasket—float needle seat
245	Valve assembly—dual capacity pump
246	Gasket-valve assembly
250	Plug—pump discharge (retainer)
251	Ball—pump discharge
252	Baffle—pump well
255	Primary metering jet assembly
265	Mixture control/dual capacity pump solenoid assembly
320	Vacuum break assembly—secondary side (rear)
321	Screw—secondary side (rear) vacuum break assembly attaching

322	Link—secondary side (rear) vacuum break to choke
330	Rivet—choke cover attaching
331	Retainer—choke cover
335	Electric choke cover and stat assembly
340	Choke housing assembly
341	Screw and washer assembly—choke housing to float bowl
345	Screw—choke stat lever attaching
348	Lever—choke stat
350	Intermediate choke shaft, lever and link assembly
352	Fast idle cam assembly
354	Lever—intermediate choke
356	Link—choke
360	Lever—secondary throttle lockout
364	Seal—intermediate choke shaft
370	Nut—fuel inlet
372	Gasket—fuel inlet
375	Filter—fuel inlet
377	Spring—fuel filter
380	Screw-throttle stop
381	Spring—throttle stop screw
400	Throttle body assembly
401	Gasket—float bowl to throttle body
405	Screw assembly—float bowl to throttle body
410	Link—pump
420	Needle—idle mixture
421	Spring—idle mixture needle
422	Plug—idle mixture needle
425	Screw—fast idle adjusting
426	Spring—fast idle adjusting screw
500	Solenoid and bracket assembly
501	Screw—bracket attaching
510	Throttle kicker assembly
511	Bracket—throttle kicker
512	Nut—throttle kicker assembly attaching
513	Washer—tab locking

Part description E4MED exploded view. List courtesy Rochester Products Division, GMC.

Remove secondary metering rods by lifting hanger from air horn. Drawing courtesy Rochester Products Division, GM.

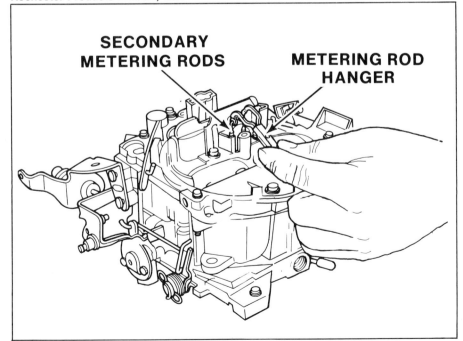

SECONDARY METERING RODS

METERING ROD HANGER

Remove screws holding vacuum-break assembly to detach from air horn.

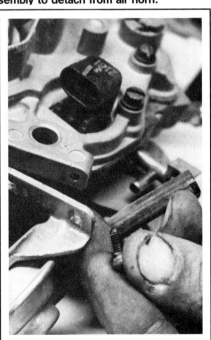

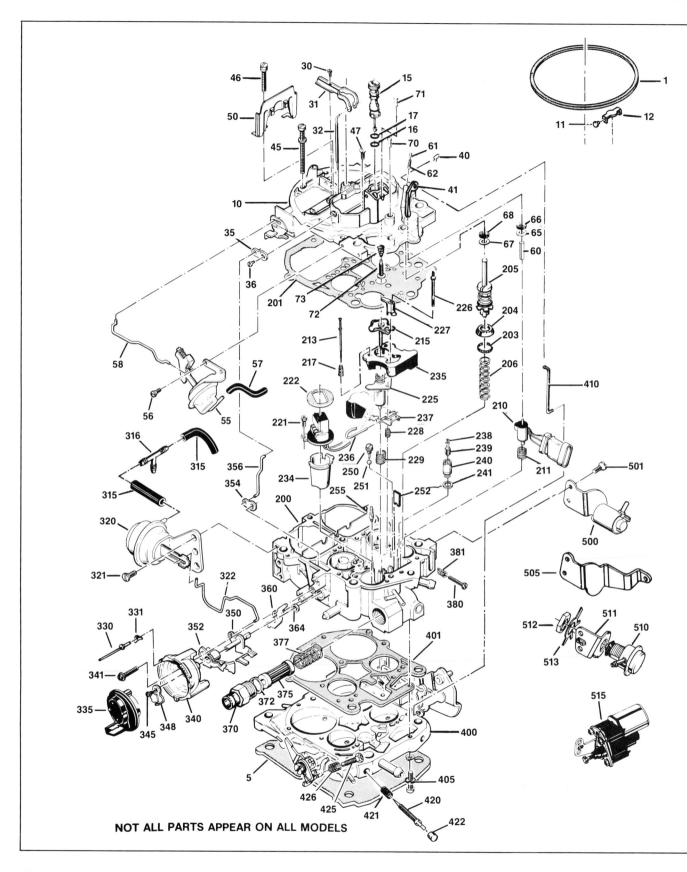

NOT ALL PARTS APPEAR ON ALL MODELS

Model E4ME

1	Gasket—air cleaner		239	Needle—float
5	Gasket—flange		240	Seat—float needle
10	Air horn assembly		241	Gasket—float needle seat
11	Rivet—cover attaching		250	Plug—pump discharge (retainer)
12	Cover—air bleed valve		251	Ball—pump discharge
15	Air bleed valve assembly		252	Baffle—pump well
16	O-ring—air bleed valve—lower		255	Primary metering jet assembly
17	O-ring—air bleed valve—upper		315	Hose—secondary side (rear) vacuum break
30	Screw—secondary metering rod holder attaching		316	Tee—secondary side (rear) vacuum break
31	Holder—secondary metering rod		320	Vacuum break assembly—secondary side (rear)
32	Rod—secondary metering		321	Screw—secondary side (rear) vacuum break assembly attaching
35	Lever—choke		322	Link—secondary side (rear) vacuum break to choke
36	Screw—choke lever attaching		330	Rivet—choke cover attaching
40	Retainer—pump link		331	Retainer—choke cover
41	Lever—pump		335	Electric choke cover and stat assembly
45	Screw assembly—air horn to throttle body		340	Choke housing assembly
46	Screw assembly—air horn to float bowl		341	Screw and washer assembly—choke housing to float bowl
47	Screw—air horn to float bowl (countersunk)		345	Screw—choke stat lever attaching
50	Baffle—air horn		348	Lever—choke stat
55	Vacuum break assembly—primary side (front)		350	Intermediate choke shaft, lever and link assembly
56	Screw—primary side (front) vacuum break assembly attaching		352	Fast idle cam assembly
57	Hose—primary side (front) vacuum break		354	Lever—intermediate choke
58	Link—primary side vacuum break—air valve lever		356	Link—choke
60	Plunger—sensor actuator		360	Lever—secondary throttle lockout
61	Plug—TPS adjusting screw		364	Seal—intermediate choke shaft
62	Screw—TPS adjusting		370	Nut—fuel inlet
65	Retainer—TPS seal		372	Gasket—fuel inlet
66	Seal—TPS plunger		375	Filter—fuel inlet
67	Retainer—pump stem seal		377	Spring—fuel filter
68	Seal—pump stem		380	Screw-throttle stop
70	Plug—solenoid adjusting screw		381	Spring—throttle stop screw
71	Plug—solenoid stop screw		400	Throttle body assembly
72	Screw—solenoid stop (rich mixture)		401	Gasket—float bowl to throttle body
73	Spring—rich authority adjusting		405	Screw assembly—float bowl to throttle body
200	Float bowl assembly		410	Link—pump
201	Gasket—air horn to float bowl		420	Needle—idle mixture
203	Spring—pump plunger		421	Spring—idle mixture needle
204	Cup—pump plunger		422	Plug—idle mixture needle
205	Pump assembly		425	Screw—fast idle adjusting
206	Spring—pump return		426	Spring—fast idle adjusting screw
210	Sensor—throttle position (TPS)		500	Solenoid and bracket assembly
211	Spring—sensor adjusting		501	Screw—bracket attaching
213	Rod—primary metering		505	Bracket—solenoid
215	Plunger—solenoid		510	Throttle kicker assembly
217	Spring—primary metering rod (E2M, E4M only)		511	Bracket—solenoid
221	Screw—solenoid connector attaching		512	Nut—throttle kicker assembly attaching
222	Gasket—solenoid connector to air horn		513	Washer—tab locking
225	Mixture control solenoid assembly		515	Idle speed control assembly
226	Screw—solenoid adjusting (lean mixture)			
227	Stop—rich limit			
228	Spring—solenoid adjusting screw			
229	Spring—solenoid return			
234	Insert—aneroid cavity			
235	Insert—float bowl			
236	Hinge pin—float			
237	Float			
238	Pull clip—float needle			

Exploded view Q-jet E4ME: It is representative of 1981—86 passenger-car Q-jets. Rich stop limit (#227) may not appear on some applications. Drawing and part description courtesy Rochester Products Division, GM.

DISASSEMBLY

All the suggestions made and many of the procedures used for servicing a non-electronic Q-jet apply to the newer units. Use due caution when handling electronic components, make sketches of linkages and keep a list detailing the order parts were removed; in short, employ any aid that helps organize the work.

Air-Horn Removal—Follow the procedure outlined earlier in the chapter. Most late-model air horns can be lifted straight up when all pieces are loosened. Be careful when removing the air horn not to damage the M/C solenoid connector, the TPS adjustment lever, and the small tubes protruding from the air horn. Gently lift the horn straight up.

One of the larger tubes—secondary pullover tube—may shake loose from the horn when it's removed. It will be standing vertically in the fuel bowl. Tap it back in place firmly, but without peening over the end. This is not a common problem, but I have seen several do this over the years.

Vacuum Break—Remove the two screws holding the break bracket. Slip the vacuum hose off one end. The vacuum-break assembly and connecting rod to the secondary air valves is removed by tilting the components and slipping the rod out of the air-valve lever.

TPS Plunger—Remove the TPS plunger by pushing it up through the seal in the air horn. Use your fingers only to prevent damage to the coated surface while removing plunger.

Lean-Mixture Plug—This must be driven out and replaced *if* M/C solenoid plunger travel is incorrect. See pages 94 and 95 for setting lean-mixture screw to establish correct M/C solenoid position and plunger travel. If plug is removed, it *must be replaced* to hold screw's setting. Plug is included in service kits.

Lean-Mixture Screw—This positions the M/C solenoid in the float bowl assembly and thus determines the solenoid plunger travel and primary metering. See pages 94 and 95 for adjusting it.

Completely remove the screw to release the M/C solenoid. Also remove the Phillips screw from the black electrical connector located in a well opposite the accelerator-pump well. *Carefully* lift the two assemblies up and out; they are connected by two wires.

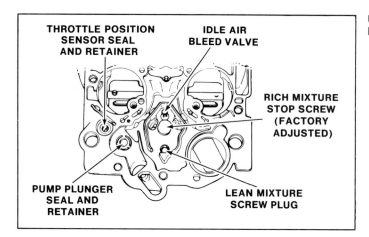

THROTTLE POSITION SENSOR SEAL AND RETAINER

IDLE AIR BLEED VALVE

RICH MIXTURE STOP SCREW (FACTORY ADJUSTED)

PUMP PLUNGER SEAL AND RETAINER

LEAN MIXTURE SCREW PLUG

Underside view of air horn. Drawing courtesy Rochester Products Division, GM.

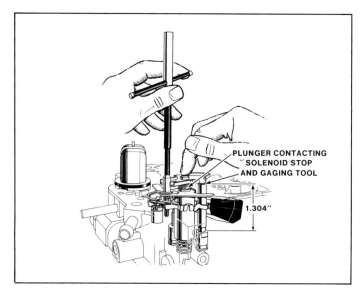

PLUNGER CONTACTING SOLENOID STOP AND GAGING TOOL

1.304"

Adjusting lean-mixture (solenoid adjusting) screw on 3- and 4-point adjustment carburetors with rich-limit stop screw in air horn:
1. Remove air horn, solenoid plunger, air horn gasket, throttle-side primary metering rod (and spring) and float bowl insert.
2. Install gage or equivalent 1.304-in. piece over primary metering jet rod guide.
3. Reinstall plunger.
4. Holding plunger down, use tool or equivalent to turn lean-mixture screw, in or out, until plunger contacts *both* solenoid stop and gage.
5. Remove plunger and gage.
6. Reinstall float bowl insert, metering rod and spring, (new) air horn gasket, plunger.
7. Remove plugs (lean- and rich- mixture) from air horn.
8. Install air horn.
9. Adjust rich-mixture (solenoid) stop screw in air horn (see nearby drawing).
10. Install new plugs to retain settings.
Drawing courtesy Rochester Products Division, GM.

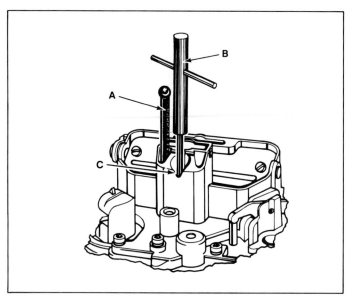

B

A

C

Adjusting rich-mixture (solenoid) stop screw on 3- and 4- point adjustment carburetors (rich mixture stop screw plug removed):
1. Insert gage or equivalent (A) through air horn vent hole to contact mixture control solenoid plunger.
2. Press (A) down and then release. Measure distance moved, which is solenoid plunger travel.
3. Insert tool or equivalent (B) in rich-mixture screw hole (C). Then, turn rich-mixture stop screw, in or out, to obtain 4/32-in. (3.175mm) plunger travel.
4. Install new lean-mixture (solenoid adjusting) screw plug and rich mixture (solenoid) stop screw plugs to retain settings.
Drawing courtesy Rochester Products Division, GM.

Top view of air horn noting: (A) lean-mixture screw plug and (B) rich-mixture screw plug.

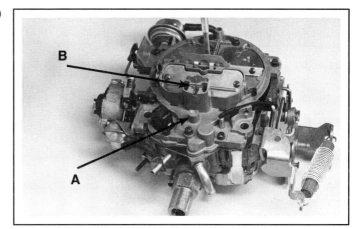

Adjusting lean-mixture (solenoid adjusting) screw on 2-point adjustment carburetors with integral rich limit stop bracket (see pages 59 and 92, #227).

1. Remove air horn, gasket, lean-mixture screw (#226), rich limit stop bracket, plunger, throttle-side primary metering rod (and spring) and float bowl insert.
2. Install gage or equivalent 1.304-in. piece (dimension C) over primary metering jet rod guide.
3. Reinstall plunger, rich limit stop bracket and lean-mixture screw.
4. Holding plunger down, use tool or equivalent (A) to turn lean-mixture screw, in or out, until plunger contacts *both* solenoid stop and gage at (B). Note tool's handle position.
5. Turn lean-mixture screw clockwise, *counting turns,* until solenoid gently bottoms.
6. Remove lean-mixture screw, rich limit stop bracket, plunger and gage.
7. Reinstall float bowl insert, metering rod and spring, plunger, rich limit stop bracket and lean-mixture screw.
8. Turn lean-mixture screw clockwise to gently bottom solenoid. Back out (counterclockwise) number of turns counted in step 5.
9. Remove lean-mixture plug from air horn.
10. Install air horn, new gasket, and new plug to retain setting.

Drawing courtesy Rochester Products Division, GM.

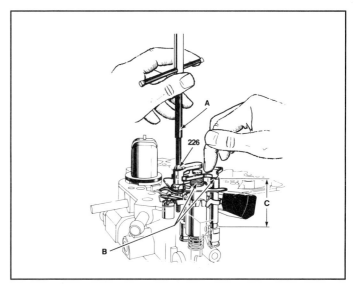

Adjusting air bleed valve:
1. Remove air bleed valve cover.
2. Insert gage or equivalent through air horn vent hole to contact solenoid plunger. Locate upper end of gage over cavity next to valve.
3. Press gage down so plunger is against solenoid stop (dimension 1.756 in.), then rotate gage toward valve.
4. Turn valve in or out until gage meets top of valve.
5. Does valve have letter identification on its top? If so: It is in a 2- or 3-point adjustment carburetor and requires no more adjusting. If not: it is in a 4-point adjustment unit and requires final adjustment with engine running. (Refer to vehicle manufacturer's idle mixture adjustment procedure.)

Drawing courtesy Rochester Products Division, GM.

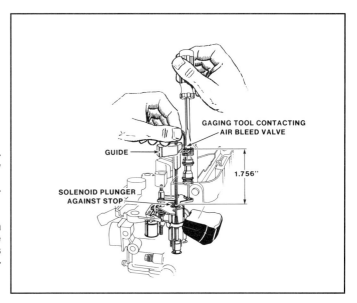

Connector plug (A) being removed. M/C solenoid (B) still in place.

Float Assembly—Lift out these parts:
- Tension spring near the float-hinge pin.
- Two primary metering rods.
- Float-hinge pin and float assembly.

You are now far enough into the carburetor to take care of nearly any metering or float problem.

Idle-Mixture Screws—Take the caps off the idle-mixture screws and remove them. *Count the turns* off the seat so they can be reinstalled at a preliminary setting.

Cleaning—Now all the metering passages are open. Spray solvent in them, let sit a short time and blow them out. An ear syringe supplies adequate pressure to check passages. Compressed air is useful for clearing passages.

Repeat this cleaning exercise two or three times if the carburetor is gummed up.

ASSEMBLY

Throttle Position Sensor—Make sure its stem is free when pushed in and let out with your finger. I have seen a few stems bent by poor service work.

Inlet Needle—Snug the inlet-needle seat with a wide screwdriver to be sure it's tight. Wipe off the inlet-needle tip and inspect it for wear or defects.

NOTE: Some people claim to have a lot of needle-and-seat problems, but I rarely find they need to be replaced.

Put the needle-and-float assembly back in place.

Float Level—Hold down the hinge pin and push down directly above the inlet needle. If you know the correct float setting, see if it's correct. It will be in the range of 5/16 in. for street use. Most service manuals supply the correct setting. Set the small tension spring back in place; be sure the larger one is in place in the bottom of the bowl.

One primary metering rod being lifted out, other still in place.

Typical float-hinge pin for early and late model Q-jets.

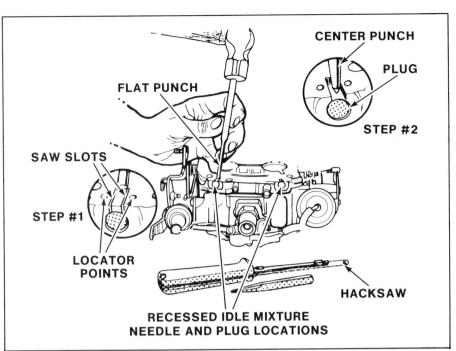

CENTER PUNCH

PLUG

STEP #2

FLAT PUNCH

SAW SLOTS

STEP #1

LOCATOR POINTS

HACKSAW

RECESSED IDLE MIXTURE
NEEDLE AND PLUG LOCATIONS

To remove idle-mixture plugs: (1) saw slots on plug edge and remove metal to reveal plug and (2) use center punch to drive plug out. Drawing courtesy Rochester Products Division, GM.

M/C solenoid plunger being inserted. TPS stem noted with pencil. Make sure they don't bind, and smoothly rise and fall in place.

One idle-mixture-screw cap knocked loose.

Make tool for setting idle mixture screws by crimping 3/16-in. steel tubing over file end to match screw head.

Idle-screw adjustment tool ready to use.

AN ASSEMBLY CAUTION

When reassembling the carburetor, pay close attention to the M/C solenoid adjustment and the TPS switch. See pages 94 and 95 for M/C solenoid adjustment procedures. As you place the air horn on the bowl, use a screwdriver or similar object to hold the TPS and accelerator pump down. I have seen a TPS ruined because a mechanic didn't take that extra measure of caution.

With the carburetor reassembled and on the car, make final adjustments to the M/C solenoid, TPS and idle-speed motor, if so equipped. Reinstall all tamperproof plugs on the carburetor. They are necessary to keep it sealed and protected from outside filth and foreign objects.

Solenoid paddle in place, lean mixture screw adjusted, connector assembly mounted: all ready for air horn fitting.

Bear, Sun, Allen are some diagnostic scopes that connect to ALDL for troubleshooting.

SERVICE EQUIPMENT TIPS

Advances in applying microprocessors to vehicle systems are just beginning. Servicing this technology requires superior diagnostic tools and, until recently, some equipment for troubleshooting an ECM had limited utility.

An ECM's message port to the outside world is the Assembly Line Diagnostic Link (ALDL). Most service oscilloscopes can tap this connector and get diagnostic information, so analysis via the ALDL is adequate in many cases. But it isn't the ideal point to gather data on transient or complex ECM problems because it allows only limited information out of the ECM and transmits it about once a second.

A new tool has been developed for extensive ECM checkout. The "Conquest", available from Digital Automotive Systems, Inc. of Garden Grove, CA bypasses the ALDL and taps directly into the ECM. Diagnostic data are read at the same rate as they are sent by the ECM (10 times per second), and a mechanic can temporarily modify ECM outputs (for example, fuel-mixture and spark-advance signals) to simulate different driving conditions and simplify troubleshooting.

Mixture Control Solenoid—Set the solenoid assembly and wire in place. Be sure the bottom of the solenoid sits on the large spring in the bottom of the float bowl. Follow the procedures on pages 94 and 95 to set the M/C solenoid metering adjustments.

Air Bleed Valve—If you change M/C solenoid plunger travel, then adjust this according to procedures on page 95.

Plastic Stuffer—Drop in the plastic stuffer over the needle and seat.

Primary Metering Rods—Put the primary metering rods in the primary jets.

Accelerator Pump—Put the accelerator pump in place. The return spring will be difficult to install.

Air-Horn Gasket—Put the air-horn gasket in place. The pump stem will fight you.

Solenoid Plunger—Drop the solenoid plunger in over the metering rods. It is designed to be guided by the vertical lean-mixture stop screw. Be sure the plunger arms meet the tops of the primary metering rods. Secure the electrical-connector assembly with the Phillips screw. Do not over-tighten.

Air Horn—Put the air horn on. Keep it flat as it's installed. Feed the pump stem through the correct air-horn hole. It should go into place easily. Do not force it. Drop in the TPS coated-plunger before securing the accelerator-pump lever. Secure the air horn, install the vacuum-break assembly and secondary metering rods.

"Conquest" bypasses ALDL and taps directly into ECM for more diagnostic information at faster update rate.

These engine factors are often overlooked when troubleshooting problems originally blamed on faulty carburetion, and neglected before selecting a replacement carburetor. Review them before investing time, effort and dollars in carburetor changes. The aggravation and money you save may be your own.

Carburetor performance is affected by:

- Spark timing.
- Valve timing.
- Temperature.
- A/F density.
- Compression ratio.
- Intake manifold design.

Early Q-jet with factory high-rise manifold on racing motor. Adapter plate at base matches throttle-bore and bolt hole spacing.

SPARK TIMING

Most vehicles have labels that specify engine idle speed and spark-advance settings. These are calculated by OEM engineers to ensure the engine combination will produce minimum emissions within specified limits. The trend is to use a retarded spark at idle and during the use of intermediate gears. Some smog systems that prevent vacuum advance in intermediate gears have a temperature-sensitive valve that allows spark advance if the engine starts to overheat and during cold operation.

Many carburetors have ports for timed spark and actual manifold vacuum. A retarded spark limits oxides of nitrogen by keeping peak pressures and temperatures at lower values than those caused by advanced spark settings. It also reduces hydrocarbon emissions. This reduces emissions—but affects economy, driveability and coolant temperature.

Fuel is being burned in the engine, but its energy is largely wasted as heat flow into the cylinder walls and as excess heat in the exhaust manifolds—which also heats the engine compartment. Consequently, the cooling system has to work harder. In essence, fuel is still burning as it passes the exhaust valve. Thermal efficiency of the engine is less because there is more wasted energy.

A retarded spark requires richer jetting in the idle and main systems to get off-idle performance and driveability. A tightrope is being walked here because the mixture must not be allowed to go lean or higher combustion temperatures and more oxides of nitrogen will be produced. If mixtures are richened too far to aid driveability, CO (carbon monoxide) emissions increase.

Because retarding the spark hurts efficiency, the throttle plate must be opened more at idle to supply enough mixture for the engine to run. This fact must be considered by the carburetor designer in positioning the off-idle slot. It must also be considered because high temperatures at idle, due to the retarded spark and high idle-speed settings, promote dieseling.

> **DIESELING**
> This describes when an engine continues to run irregularly and roughly after the ignition is off. It is aggravated by: 1) anything that remains hot to ignite A/F mixtures—such as sharp edges in the combustion chamber, and 2) fast-idle settings. The latter is most often the cause of dieseling. Poor fuel quality also contributes to this problem.

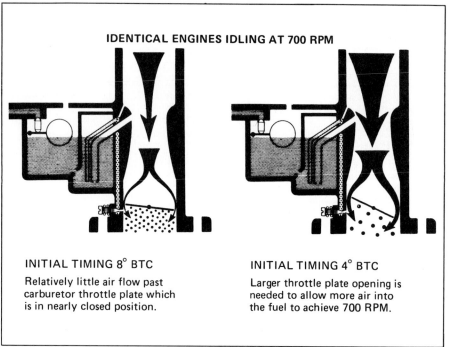

IDENTICAL ENGINES IDLING AT 700 RPM

INITIAL TIMING 8° BTC
Relatively little air flow past carburetor throttle plate which is in nearly closed position.

INITIAL TIMING 4° BTC
Larger throttle plate opening is needed to allow more air into the fuel to achieve 700 RPM.

Retarded spark requires opening throttle plate more at idle.

Distributor vacuum port (arrow) is exposed to manifold vacuum as primary throttle blades open. When blades close, there is no vacuum advance at idle.

On carburetors equipped with timed spark advance—no advance at closed throttle—a port in the throttle bore is exposed to vacuum as the throttle plate moves past the port—usually slightly off idle. The spark-advance vacuum versus airflow calibration is closely monitored in the production of the carburetor because it has a substantial effect on emissions. The distributor advance, once held to be so important for economy, has now become an essential part of reducing emissions.

VALVE TIMING

Valve timing has the most effect on idling and low-speed performance. Adding valve overlap and lift with a racing cam allows the engine to breathe better at high rpm. But it worsens manifold vacuum at idle and low speeds—creating distribution and vaporization problems.

The engine becomes hard to start, idles roughly—or not at all—has a bad flat spot coming off idle, and has very poor pulling power (torque) at low rpm. The conditions are especially noticeable when a racing cam is matched with a lean, emission-type carburetor.

Because manifold vacuum is reduced, the signal available to pull fuel mixture through the idle system is also reduced—the mixture is leaner. The throttle must be opened farther than usual to get enough mixture into the engine for idling. This can place the off-idle slot/port in the wrong relationship to the throttle. Now there is insufficient off-idle fuel to carry the engine until main-system flow begins.

When manifold vacuum is reduced, the power valve may operate, or at least open and close, because vacuum varies wildly. This is not a valid reason to remove the power assembly, but it may require altering the valve spring. This will help close the valve at lower vacuum and stabilize it.

A racing cam may magnify these problems to the point where a street vehicle becomes undriveable for anything except competition. This is especially true if the carburetion capacity has been increased to match the better breathing of the cam.

There have been cases when a racing cam was installed at the same time a carburetor was changed. If the mechanic did not understand what was occurring in the engine, the carburetor was often blamed for poor idling and the bad flat spot at off-idle. The real culprit was the racing camshaft! You'll find more information on tuning an engine with a racing camshaft on page 131.

TEMPERATURE

Temperature influences carburetion. It affects mixture ratio because air becomes less dense as temperature increases. Reduced density reduces engine power. To maintain a correct mixture ratio, make main-jet corrections of approximately one jet size smaller for every 50F (10C) ambient temperature increase.

Maximum power production requires the inlet charge be as cold as possible. For this reason, racing-engine intake manifolds are not heated. Street vehicles, on the other hand, use a heated intake manifold because the warmer mixture—although not ideal for maximum power—helps driveability.

A warm-air inlet to the carburetor or a water-heated intake manifold aids vaporization. Good vaporization ensures that the mixture will be evenly distributed to the cylinders.

ICING

Icing occurs most frequently at 40F (4C) and a relative humidity of 90% or higher. It is usually a problem at idle, with ice forming between the throttle plate and bore. It happens when the vehicle has been run a short distance and stopped with the engine idling. Ice builds up around the throttle plate and shuts off the mixture flow—the engine stops.

Once the engine stalls, no vaporization is occurring, so the ice promptly melts and the engine can be restarted. This sequence takes place until the carburetor body is hot enough and vaporization does not cause icing.

"Turnpike Icing"—This is icing in the venturi. It can happen when driving at constant speed for extended periods in icing conditions. In this case, ice build-up restricts venturi size so the engine runs slower and slower.

Vaporization of fuel removes heat from the surrounding parts of the carburetor, hence there is a greater tendency for this to occur where vaporization is best, namely, in small venturis.

Icing is not a major problem in modern vehicles because they have exhaust-manifold stoves to warm the air to the carburetor. Thermostatic valves on some engines shut off the hot-air flow when underhood temperatures reach a certain level. On some high-performance engines, a vacuum diaphragm opens the air cleaner to a hood scoop or other cold-air source at low vacuums and heavy loads.

HEAT

Percolation—At the other end of the thermometer we find a problem called *percolation*. It usually occurs when the engine is stopped during hot weather or after it is fully warm. Engineers call these conditions a *hot soak*. No cooling air is being blown over the engine by the fan or vehicle motion. Heat stored in the engine block and exhaust manifolds is radiated and conducted into the carburetor, fuel lines and fuel pump.

Fuel in the main system between the fuel bowl and main discharge nozzle can boil or percolate, and the vapor bubbles push or lift liquid fuel out of the main system into the venturi. The fuel falls onto the throttle plate and trickles into the manifold. Excess vapor from the fuel bowl—and from the bubbles escaping from the main well—are heavier than air, so they drift down into the manifold.

The extra vapor and fuel make the engine difficult to start when hot, so a long cranking period is usually required. In severe cases, enough fuel collects in the manifold to run into cylinders with open intake valves. The fuel washes the oil off cylinder walls and rings and shortens engine life.

Percolation is aggravated by fuel boiling in the fuel pump and the fuel line to the carburetor. This creates fuel-supply pressure as high as 11—15 psi—sufficient to force the inlet-valve needle off its seat. Fuel vapor and liquid fuel are forced into the bowl so the fuel level is raised. So, it is that much easier for vapor bubbles to lift fuel to the spillover point.

Percolation problems are resolved with several solutions. The main system is designed so vapor bubbles lifting fuel toward the discharge nozzle tend to break before they push fuel out of it. Fuel levels are carefully established to provide as much lift as can be tolerated. In some instances, fuel will be made to pass through an enlarged section at the top of the main well or standpipe to discourage vapor-induced spillover.

Gaskets and insulating spacers are used between the manifold and carburetor, and between the carburetor base (throttle body) and fuel bowl. In some instances, an aluminum heat deflector or shield is used to keep some of the engine heat away from the carburetor.

Another method to reduce percolation and hot-starting problems is with internal bleeds in the fuel pump or a vapor-return line on the carburetor. When the bleed and return line are used, any pressure build-up in the fuel line escapes into the fuel tank or fuel-supply line.

Vapor Lock—Another problem caused

Heat from exhaust-manifold stove is piped through flex duct to air cleaner's inlet snorkel.

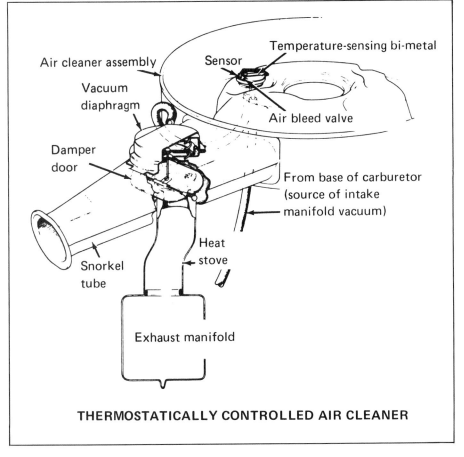

Air cleaner assembly

Sensor

Temperature-sensing bi-metal

Vacuum diaphragm

Air bleed valve

Damper door

From base of carburetor (source of intake manifold vacuum)

Snorkel tube

Heat stove

Exhaust manifold

THERMOSTATICALLY CONTROLLED AIR CLEANER

by high temperature is boiling of fuel in the fuel line between the fuel pump and carburetor. Fuel can even boil in the pump itself. When the fuel pump and line are filled with hot fuel, the pump can then supply only a *mixture* of vapor and liquid fuel to the carburetor.

Because little *liquid* fuel is delivered when accelerating after a hot soak, the carburetor fuel level drops and the mixture is leaned. In fact, the bowl may be nearly emptied, partially exposing the jets. When they are exposed, the carburetor can't meter a correct mixture because jets are designed to flow liquid fuel—not a combination of liquid and vapor. This lean condition is called *vapor lock*.

Rochester designers combat vapor lock by isolating fuel-metering components from hot metal. With the 2G, 4G and H-series carburetors the idle tubes, idle-fuel channels, nozzles and main fuel-aspirator channels are contained in a cluster unit. These are held in place on a gasket by screws.

The Monojet and Q-jet have a thick insulating gasket between the throttle body and fuel bowl to prevent excessive heat transfer to the metering orifices and channels. In some cases the gasket between the throttle body and manifold is extra thick so less heat will be transferred from the engine into the carburetor.

In extreme cases, fuel lines may have to be routed to keep them away from heat, such as from the exhaust system. If they can't be relocated, it is usually possible to insulate them. Fuel-cooling cans also help, as do electric fuel pumps designed to *push* fuel to the carburetor.

The two-barrel cluster tends to keep some of metering components cooler. It includes emulsion tubes, idle tubes, main and idle air bleeds and main nozzles.

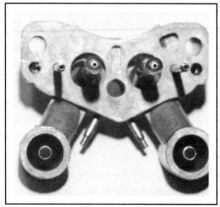

DENSITY

In Engine Air/Fuel Requirements, page 23, engine volumetric efficiency was related to A/F mixture density—the higher the density, the more HP.

Mixture density depends on atmospheric pressure, which varies with altitude, temperature and weather conditions, and it is also affected by intake-system layout.

Density is increased when the carburetor is supplied with cool air and when the intake manifold is not heated. Density is reduced if the inlet air is heated or if the mixture is heated as it enters the manifold. It continues to be reduced as it picks up heat from the manifold and cylinder-head passages, hot valves, cylinder walls and piston heads.

Although density affects carburetor capacity, its greatest effect is on mixture. Primary density changes are caused by changes in altitude. A standard rule of thumb is to reduce jet size by 1% flow for each 1000-ft. increase.

COMPRESSION RATIO—CR

HIGH COMPRESSION

High compression improves engine performance by increasing the burning rate of the mixture; then peak pressures and peak torque can approach the engine's maximum. Unfortunately, high compression also increases the emission of hydrocarbons (HC) and oxides of nitrogen (NOx).

Until about 1970, high-compression engines with up to 11:1 CR were available in high-performance cars. By 1971, manufacturers were reducing CR, and in 1972 most cars had no more than 8.5:1 or 8:1 ratios. Reducing compression slows the burning rate of the mixture, and peak pressures that encourage NOx production can't be reached. Reduced compression also increases the amount of heat transferred into cylinder walls because more combustion happens as the piston is descending, thereby raising the exhaust temperature.

LOW COMPRESSION

Lowering CR reduces HC emissions by reducing the surface-to-volume ratio of the combustion chamber. The greater the surface-to-volume ratio, the more surface cooling and increased HC concentration.

Using low compression increases the

A/F requirement at idle because more residual exhaust gas remains in the clearance volume and combustion chamber when the intake valve opens—the mixture is excessively diluted. This causes off-idle driveability problems. In effect, low compression accomplishes some degree of *exhaust gas recirculation (EGR)* without plumbing or hardware. See Emissions Control, page 166, for more details on this hardware.

Raising or lowering the CR usually doesn't affect main-system fuel requirements, so significant jet changes are not required when this modification is made. Raising compression may require slightly less ignition advance in some cases.

COMPRESSION & FUEL OCTANE

An engine's compression ratio and fuel octane requirement are related. As compression is increased, higher octane fuel must be used to avoid detonation and preignition—often referred to as *knock*. Similarly, lowering an engine's compression reduces its octane requirement.

After World War II, engines were designed and produced with high compression ratios to obtain better efficiency. By 1969, some had 11:1 compression ratios, and 100+ octane leaded pump fuel was readily available.

During the '70s, stiffer emission controls brought about the end of such high ratios and high-octane leaded fuel. In the mid-'80s, production-engine ratios reversed direction. They reached 10:1 as high-octane unleaded fuel became readily available and electronic engine controls—knock sensors for one—were used with superior emission control designs.

INTAKE MANIFOLD

The intake manifold is a mount for the carburetor; it connects carburetor throats and cylinder-head ports. Sometimes a manifold is a mount for other accessory devices, serves as tappet-chamber cover, and can provide coolant passages or attachments. The manifold divides air and fuel equally among the cylinders at all speeds and loads. Ideally, each cylinder receives the same amount of mixture with the same A/F ratio.

This ideal distribution is usually not achieved with OEM production manifolds.

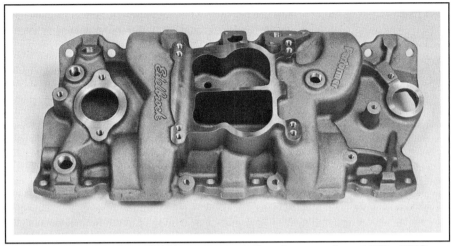

Edelbrock "Performer" manifold improves overall performance from 1500—5500 rpm while maintaining emissions and driveability. Photo courtesy Edelbrock.

Manufacturers get cylinder distribution close to the ideal during most driving ranges. Aftermarket producers try to improve on OEM designs, but for most street applications they seldom perform miracles, either.

DESIGN

Manifold passages should have approximately equal length, cross-sectional area and geometric arrangement. For various reasons, this cannot always be done, so flow-equalizing features are sometimes used to make unequal-length passages perform as well as possible.

Because the manifold distributes gasoline in mixture and liquid form, some liquid gas is usually moving around on the walls and floor of the manifold. Controlling this liquid gas by devices such as sumps, ribs and dams is important in trying to attain equal A/F ratios for all cylinders. As an aid to vaporization and as part of controlling unvaporized fuel—or that which has dropped out of the mixture due to an increase in pressure—heating the manifold is essential in all except racing applications. To avoid gravity force that could cause uneven distribution of the liquid fuel, the manifold floor and carburetor base are parallel to the ground—instead of being parallel to the crankshaft centerline, which may be angled for drive-train alignment.

Small venturis aid A/F velocity through the carburetor and manifold, and promote vaporization and distribution. High-speed airflow through the carburetor and manifold maintains turbulence. Velocity and turbulence keep fuel droplets in suspension, thereby tending to equalize cylinder-to-cylinder A/F ratios. If the mixture is allowed to slow, as happens when passage size is increased, fuel may separate from the air and deposit on the manifold surfaces. Variations in the A/F distribution result.

Passage design should be such that mixture flow is fast enough to sustain low and mid-range throttle response without reducing volumetric efficiency at high rpm. The internal design of the passages must carry mixture without forcing the fuel to separate from the air. When the A/F mixture is forced to turn or bend, the air can turn quicker than the fuel and they may separate. When fuel separates, a cylinder being fed with the mixture after the bend will receive a different A/F ratio.

Pulsing within the manifold must also be accommodated. This condition, also called *back-flow* or *reversion*, occurs twice during a four-cycle sequence. A pulse of residual exhaust gas enters the intake manifold during the valve-overlap period when both intake and exhaust valves are off their seats. Another pulse toward the direction of the carburetor occurs when the intake valve closes. Either of these pulses may hinder flow in the manifold. In the most severe cases, the pulsing travels through the carburetor toward the atmosphere, and appears as a fuel cloud standing above the carburetor inlet. This is called *standoff*.

The carburetor meters fuel into the air stream *regardless of the direction of the stream*—either down through the carburetor or up through it from the manifold. This bi-directional metering occurs to some degree in most manifolds and in a marked degree in others.

In many performance applications, the A/F ratio differences between cylinders are remedied by using unequal jet sizes. Richer (larger) jets are used in a section of the carburetor feeding lean cylinders, and leaner (smaller) jets are used for those carburetor barrels feeding rich cylinders. Although, unequal jetting—commonly called *cross- or staggered-jetting*—is a partially effective remedy. In such cases, the manifold is the primary problem.

MANIFOLD TYPES

There are a number of different manifold types:

- Single-plane.
- Two-plane.
- High-rise single- or two-plane.
- Individual or isolated-runner (IR).
- IR with plenum chamber (Tunnel Ram).

Single-Plane—In a *Car Life* article entitled "Intake Manifolding," Roger Huntington said, "The very simplest possible intake-manifold layout would be a single chamber that feeds to the valve ports on one side and draws from one or more carburetor venturis on the other side. This is called a *common chamber* or *runner (single-plane)* manifold. Common-chamber manifolds have been designed for all types of engines—in-line sixes and eights, fours and V8s. It's the easiest and cheapest way to do the job. In fact, most current in-line four- and six-cylinder engines use this type of manifold."

Huntington also wrote, "But we run into problems as we increase the number of cylinders. With eight cylinders there is a suction stroke starting every 90° of crankshaft rotation. They overlap. This means that one cylinder will tend to rob mixture from the one immediately following it in the firing order, if they are located close to each other on the block. With a conventional V8 engine there are alternate-firing cylinders that are actually adjacent—either 5-7 on the left in AMC, Chrysler, GM firing order, or 5-6 or 7-8 in the two Ford firing orders."

Two-plane (Cross-H) manifolds have been shown to be more throttle- and torque-responsive at low- and mid-range than most, but not all, single-plane designs. In the early '70s, new single-plane manifolds were introduced with high mixture-stream speed, good fundamental A/F distribution, and adequate inlet flow at high rpm to permit strong running at both low and high engine speeds. Examples included Edelbrock's *Tarantula* and *Torker* and Weiand's *X-terminator* and *X-celerator*.

At first glance, dual-port manifolds may appear to be two-plane designs, but they are two single-plane manifolds. A small- and large-passage manifold are stacked in a single casting. The network of small passages connected to the primary side of the carburetor provides high-speed mixture flow for good distribution and throttle response at low- and mid-range rpm. Larger passages connected to the secondary portion of the carburetor supply the extra capacity required for high-rpm operation.

Some experimental dual-port, *two-plane* manifolds, were developed by automobile manufacturers in the past. The high cost of production offset their limited value in most applications.

Carburetor selection is explained later in this chapter, but note here that smaller carburetors can be used effectively with single-plane manifolds because the common chamber damps out most pulsing, which reduces flow capability. A small carburetor tends to quicken mixture flow and thereby improves throttle response at the low- and mid-range.

Two-Plane—These are simply two single-plane manifolds arranged so each is fed from one half (one side) of a two- or four-barrel carburetor—and each of the halves feeds one half of the engine. A Cross-H manifold for a V8 is a typical example. Each half of the carburetor is isolated from the other—and from the other half of the manifold—by a plenum divider.

The manifold passages are arranged so successive cylinders in the firing order draw first from one plane—then the other. In a 1-8-4-3-6-5-7-2 firing order, cylinders 1,4,6,7 draw from one manifold plane and one half of the carburetor. Cylinders 8,3,5,2 are supplied from the other plane and side of the carburetor.

Because there is less air mass to activate during each inlet pulse, throttle response is crisper and mid-range torque is improved.

V-8 MANIFOLD TYPES

CROSS-H

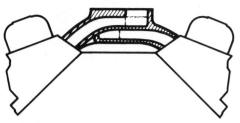

Cross-H or two-level manifold feeds half of cylinders from one side of carburetor — other half from other side of carburetor. Two sides of manifold are not connected.

SINGLE-PLANE

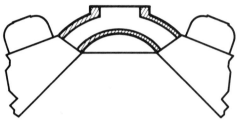

Single-plane manifold has all cylinder intake ports connected to a common chamber fed by the carburetor.

PLENUM-RAM

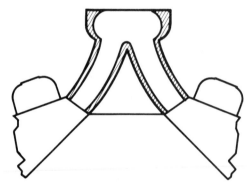

Plenum-ram manifold has a plenum chamber between passages to intake ports and carburetor/s.

ISOLATED-RUNNER

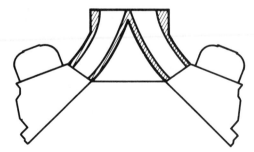

Isolated-runner (IR) manifold uses an individual throttle bore of a carburetor for each cylinder. There is no interconnection between the intake ports or throttle bores.

Early Q-jet on Edelbrock "Tarantula" manifold: Popular combination for drag racing.

Cross-H design shown in cutaway. Left side of carburetor supplies lower manifold plane (arrows). Right side supplies upper plane.

Particularly when it's compared with a conventional single-plane manifold.

Dividing the manifold in two sections may cause flow restriction at high rpm because only one half of the carburetor flow capacity and manifold volume are available for any intake stroke. Thus, the divider is sometimes reduced in height—or removed completely—to make more carburetor and manifold capacity available. This sacrifices bottom-end performance, while top-rpm capability is increased, because the mixture speed is reduced at low rpm, due to the volume increase on each side or section.

On engines built for high-rpm operation, removing dividers can sometimes be a tuning plus. For street and track applications where the camshaft and other engine parts are chosen to build torque into an engine, the divider should be left in place. Small amounts of divider can be removed until optimum divider height for a particular engine/parts combination and application is reached.

High-Rise or High-Riser?—Optional high-rise, high-performance manifolds have been offered by some manufacturers and aftermarket manifold producers. High-rise should not be confused with *high-riser,* which refers to spacing-up the carburetor base so a longer riser is available to straighten flow before it enters the manifold. The function of a high-riser is to improve distribution by eliminating directional effects caused by a partially opened throttle.

A *high-rise* manifold aligns the cylinder-head port angle with that of the manifold passage or runner. In a V8 engine, this usually means that the entire network of manifold runners is raised—along with the carburetor mounting.

Both high-rise and high-riser designs may be used with either single-plane or two-plane manifolds.

Individual-Runner (IR)—Individual- or isolated-runner (IR) manifolds are racing manifolds using one carburetor throat and one manifold runner per cylinder-head inlet port. Each of these runner/carburetor-throat arrangements is isolated from its neighboring cylinders.

Carburetors used in IR setups should have complete fuel-metering for each carburetor barrel because there is no way the cylinders can share metering. The Q-jet doesn't lend itself to this type of manifold because of its small primary bores. An important benefit of an IR manifold is it allows tuning to take advantage of ram effect. Shorter tuned lengths cause the torque peak to occur at higher rpm.

In IR systems, carburetors must be larger than those needed for other manifold types because each cylinder is being fed by just one carburetor throat or barrel. Another reason for needing a larger carburetor is that the severe pulsing existing in such systems tends to reduce carburetor flow capacity in the rpm range where standoff becomes severe. In any IR system, some method of standoff containment should be used. The usual method is a stack long enough to contain the standoff atop the carburetor inlet.

Plenum-Ram Manifold—This manifold uses a plenum chamber between the carburetor base and manifold runners. This chamber helps dissipate the strong pulsing so less of it enters the carburetor to disrupt flow. And, it allows cylinders to share carburetor flow capacity. In the typical dual-quad plenum manifold, three or four cylinders will be drawing mixture from the plenum, which is being fed by eight carburetor bores.

MULTIPLE-CARBURETOR MANIFOLDS

Until about 1967, two or more car-buretors were considered essential for a modified or high-performance engine. No self-respecting automotive enthusiast would consider building an engine with less than two carburetors. The present-day trend is to use a single four-barrel carburetor for both street and competition applications. Sophisticated four-barrel carburetors with small primaries and large, progressively operated secondaries have advanced the construction of single-carburetor manifolds that provide excellent performance and low emissions.

Multiple carburetion is useful for extended high-rpm driving because it reduces the pressure drop across the carburetors to an absolute minimum—there will be the least possible HP loss from this restriction. Of course, low-rpm operation is not a consideration for such driving, so the extra venturi area or carburetor flow capacity does not create any problems.

Many engines have been sold with two- and three-carburetor manifolds as original stock equipment. But by 1972, all of the manufacturers except Chrysler had converted their high-performance engines to single four-barrel configurations.

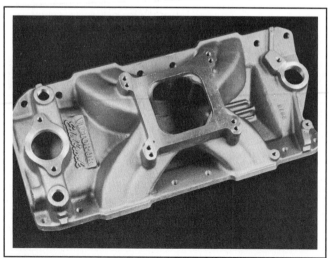

No center divider on Edelbrock "Victor Jr." increases performance from 3500-7500 rpm. Photo courtesy Edelbrock.

Ford Cobra Jet 428 used Q-jet as original equipment on stock manifold.

Should you be the proud owner of a car with 2 X 4s or 3 X 2s, avoid buying new mechanical linkages to open the throttles simultaneously. Throughout this book the need for adequate mixture velocity is stressed, and suggestions are made for selecting the correct carburetor capacity for engine size and operating rpm. Simultaneous opening of multiple throttles—except for individual runner manifolds—goes against some of these recommendations. The vehicle is hard to drive because the velocity through the carburetors at low and medium speeds is reduced, when compared with more desirable, progressively operated secondaries.

CARBURETOR SELECTION

Many people assume a bigger carburetor is a better carburetor. My experience suggests otherwise. For street applications and general driving, a bigger carburetor is often worse. The temptation to replace a stock carburetor is powerful, and is reinforced by rumor and advertising. But the carburetor is only one variable in a performance vehicle's equation.

First, think about the *combination* of axle ratio, tire diameter and engine when selecting a carburetor. Suppose you are building or buying a vehicle primarily for around-town driving. If fuel economy is important, consider a compact vehicle with a four-cylinder, or small V6 engine. It would be wasteful to buy gearing and carburetion for higher speed driving. A *small* carburetor may handle nearly all the in-town driving requirements.

If racing is the goal and the engine will operate at 6000 rpm and higher, then size the carburetor accordingly to fit the engine size and rpm capability—considering camshaft and gear ratio. Be prepared to sacrifice economy and driveability during any attempt to use this vehicle on the streets or highways.

Optimize the combination for a vehicle's intended use. Make all elements fit: cam, carburetor, manifold, gear ratio, compression, ignition and exhaust. Just changing the carburetor is likely to degrade performance and vehicle utility.

IMPROVED BREATHING INCREASES POWER

The stock passenger-car engine is a reasonably efficient air pump over a fairly wide rpm range—from low rpm to 5,000 or slightly higher. It has a reasonably flat torque curve throughout its operating

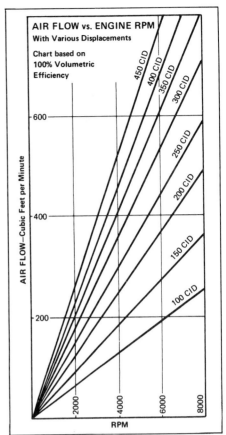

Use chart to determine carburetor airflow requirement. Correct for VE of engine. See Chapter 2 for more information.

Q-jet with velocity stack on Weiand high ram manifold.

1957 Oldsmobile J-2 "Rocket" engine with center 2GC (with choke) and vacuum-activated front and rear 2G carburetors.

1966 Olds 442 with center 2GC (with choke) and end 2Gs. This setup had mechanical linkage to open end carburetors when center one was 65% open.

Rochester Carburetor Flow Ratings	CFM
Quadrajet: Air flow at 90° air-valve opening	
1-3/32-in. venturi (primary)	750
1-7/32-in. venturi (primary)	800
Dualjet	
1-3/32-in. venturi	227
1-7/32-in. venturi	287
Varajet (staged 2 barrel)	
28mm primary	375
30mm primary	397
Model 2G: 1-1/4-in. flange, 1-7/16-in. throttle bore,	
1-3/32-in. venturi	278
Model 2G: 1-1/2-in. flange, 1-11/16-in. throttle bore	
1-3/16-in. venturi	352
1-1/4-in. venturi	381
1-5/16-in. venturi	423
1-3/8-in. venturi	435
Model 4G: Throttle bore & venturi size in inches	
1-7/16—1-1/8 primary; 1-7/16—1-1/4 secondary	486
1-7/16—1-1/8 primary; 1-11/16—1-15/32 secondary	553
1-9/16—1-1/8 primary; 1-11/16—1-15/32 secondary	692
Monojet	
1-7/16-in. throttle bore, 1-7/32-in. venturi	160
1-11/16-in. throttle bore, 1-5/16-in. venturi	210
1-11/16-in. throttle bore, 1-1/2-in. venturi	250
NOTE: 4 barrels rated at 1.5-in.Hg pressure drop.	
2 barrels rated at 3.0-in.Hg pressure drop.	
1 barrels rated at 3.0-in.Hg pressure drop.	

range. Its pumping capabilities can be optimized for better pumping within a narrower rpm band by improving breathing.

Reducing restrictions increases charge density to the cylinders. Improvements are typically made by changes to the carburetor (higher capacity), intake tract (manifold through the ports and valves), exhaust system (headers or free-breathing mufflers), and valve timing (camshaft). Such changes are similar to those a pump designer would do to make an efficient pump *at a particular rpm*. But, there is a trade-off. When you make the engine a better pump, the torque peak and the entire torque curve are lifted to a higher rpm band. Usually, low rpm performance is *reduced*.

Not all engines can be improved for higher performance (better breathing) without extensive modifications. This is because the designers have purposely optimized the low rpm performance with restrictions such as: small-venturied carburetors, tiny intake-manifold passages and ports, small valves actuated by short-duration camshafts with lazy action, low compression, combustion chamber design and restricted exhaust systems. Truck engines, low-performance passenger-car engines and industrial engines are good examples of this optimization.

Keep in mind these design restrictions when selecting a carburetor, because it is difficult—if not impossible—to upgrade the performance of such an engine. As they say, "You can't make a race horse out of a plow horse." If the engine is so restricted that it produces peak power at 4,000 rpm, selecting a carburetor for feeding the same engine at 6,000 to 7,000 rpm is not wise. It's unlikely that the engine will ever run at those speeds—at least, not without extensive modifications. And, a too-large carburetor will definitely worsen the performance that was previously available. It's rare when a transportation vehicle will perform its duties—performance, economy and driveability—better with a replacement carburetor. With a bigger one it's even less likely.

CHOOSING CARBURETOR CAPACITY

The larger the carburetor, the higher the airflow must be before the main systems begin to feed. If the carburetor is too large, the pump shot will be consumed before the main systems start. This results in a sag or bog. The Q-jet provides a supplement of fuel to handle this transition period when fuel is needed to cover sudden throttle openings.

On most units, two holes are placed above or below the leading edge of the secondary air valves. Those holes are fuel feeds that respond to high depressions that exist under the air valve when the huge secondary throttle blades are opened. During the first few degrees of air-valve opening, the holes are uncovered and they respond with a spray of fuel that temporarily fulfills the demand until the nozzle has time to respond. This is RPD's way of handling this requirement, as opposed to a second pump. Many enthusiasts and mechanics are unaware this secondary system exists on the Q-jet.

INSTALLATION PRELIMINARIES

Check the screws attaching the top of the carburetor to the throttle body. Gaskets will compress after they have been installed, so it's important to check the screws so no air or fuel leaks after installation. If you're disassembling a new carburetor to inspect its internals, assemble it immediately afterward so the gaskets will not shrink into an unusable condition.

Some people advocate taking the throttle butterflies off of their shafts or loosening the screws so throttles can be centered in the bores. *Don't do this* unless a serious binding problem is apparent.

The throttle levers contact adjustable stops, and the carburetor is checked for correct airflow before it leaves the factory.

Clearances around the throttles are factory-set. If you break off one of these screws in the throttle shaft, you have a time-consuming correction to a self-inflicted task waiting.

In general, plan to bolt the carburetor on and run it before making any tuning changes. Start with what RPD engineers have found to work successfully. They produce hundreds of thousands of carburetors every year—for all kinds of applications—and chances are good that the carburetor will be close to correct when it is bolted on the engine. This is particularly true if using a 1981-or-later unit—there is not much to alter.

INSTALLING A CARBURETOR

See the Quadrajet Service chapter for instructions on removing and installing a carburetor. Although written for that particular application, the information applies when installing most carburetors.

MATCH CARBURETOR TO MANIFOLD
Carburetor selection may be limited by the type of manifold available for the engine. Smaller carburetors can usually be installed with adapter flanges. Larger carburetors shouldn't—in most cases—be mounted on a manifold designed for a smaller carburetor, *unless* the manifold can be modified to eliminate restrictions at the carburetor-to-manifold flange.

DETECTING MANIFOLD LEAKS
An instant indicator of a manifold leak is a rougher-than-normal idle with the throttle at the factory-set curb-idle position. You may hear a hissing or whistling sound. First, check all hoses attached to the carburetor base or to the intake manifold. Make sure that no hoses are cracked or broken. Take special care to check the underside or hidden portions of the hose where a leak might not be obvious with just a glance. Also make sure that all manifold-vacuum ports have plugs installed where required.

Some mechanics squirt a little solvent on hoses or the manifold to locate leaks in manifold gasketing. When combustible solvent enters through a leak, the idle speed increases noticeably. *Use extreme care if applying this detection method* because of the clear fire hazard of squirting combustible solvent droplets on a running engine's manifold. It is not a foolproof diagnostic procedure either. Solvent vapors can be sucked into the air cleaner or carburetor air horn, and engine speed will increase from these, misleading you into thinking you've found a vacuum leak.

Q-jet primary venturi size for 800 cfm unit is 1-7/32-in. (top) and 1-3/32-in. for 750 cfm (below).

Optimize component combinations, not just carburetor. Edelbrock "Performer" manifold and matching "Performer Plus" camshaft and lifters for Buick 231 CID V6. Photo courtesy Edelbrock.

7

High Performance Carburetion

Super Stock Camaro record holder racing with Q-jet and Edelbrock manifold combination.

NHRA Super Stock Chevy with Q-jet on Edelbrock Tarantula manifold.

This chapter introduces concepts and conditions that affect the ultimate performance obtainable from an engine and carburetor combination. Then I give some advice on how to plan your tuning efforts so you won't waste time and money. Performance and tuning modifications for the Q-jet and how to apply it to different off-road driving are then specified.

CARBURETOR RESTRICTION

Carburetors are tested at a given pressure drop at WOT to obtain a cfm rating indicative of flow capacity. One- and two-barrel carburetors are tested at 3.0-in.Hg pressure drop. Four-barrel carburetors are tested at 1.5-in.Hg pressure drop.

To compare the two ratings, use these equations:

$$\frac{\text{Equivalent flow}}{\text{at 1.5 in.Hg}} = \frac{\text{CFM at 3.0 in.Hg}}{1.414}$$

$$\frac{\text{Equivalent flow}}{\text{at 3.0 in.Hg}} = \text{CFM at 1.5 in.Hg} \times 1.414$$

The one- and two-barrel rating was adopted because low-performance engines typically showed a WOT manifold-vacuum reading of 3.0 in.Hg. When four-barrel carburetors and high-performance engines became commonplace, they were rated at 1.5-in.Hg pressure drop for two reasons.

First, this rating was close to the WOT manifold vacuum being measured in these engines. Second, most of the carburetor testing equipment had been designed for smaller carburetors. The pump capacity on this test equipment was not adequate to provide 3.0-in.Hg pressure drop through larger carburetors. The 1.5-in.Hg drop was about the limit of the pump capacity.

For maximum output, it is essential to have the carburetor as large as possible—*consistent with the required driving range.* Using a lot of flow capacity can reduce inlet-system restriction and increase volumetric efficiency at WOT and high rpm. Such carburetion arrangement can compete with fuel injection in terms of performance because the restrictions are minor. However, it should be noted that the dual-quad installations typically used by professional racers are not capable or providing usable low- or mid-range performance. These engines are typically operated in a narrow range of 6000—8500 rpm. Such installations have not usually used Rochester carburetors.

Dual quads or multiple carburetor configurations are for high rpm use only. This racing setup uses four Rochester two barrels.

AIR CLEANER

Because every engine needs an air cleaner to limit expensive cylinder wear caused by dirt, it makes sense to use one that will not restrict the carburetor's airflow capabilities. By avoiding restrictions in the air cleaner, you allow your engine to develop full power.

One application in which an engine might be run without an air cleaner would be on drag-car or drag-boat engines being operated where there is no dust in the air or pit—an unlikely case to say the least. Even then, when an air cleaner is removed from the carburetor, the air cleaner base should be kept because its shape ensures an efficient entry path for the incoming air and helps to keep it from being heated by the engine.

The accompanying table shows the results of tests that dispel some of the common fallacies about air cleaners and their capabilities. In general, a tall, open-element air cleaner has the least restriction.

Air Cleaner Comparison	
Cleaner Type	**WOT Air Flow (CFM)**
None	713
Chevrolet 396 closed-element with single snorkel	480
Chevrolet 396 closed-element with single snorkel cut off at housing	515
Same as above, but with two elements	690
Chevrolet high-performance open-element unit	675
Same as above, but with two elements	713
14-inch diameter open-element accessory-type air cleaner	675
Chevrolet truck-type element (tall) used with accessory-type base & lid	713
Foam-type cleaner (domed flat-funnel type)	675

NOTE: All data obtained with same carburetor. New, clean paper elements used in all cases except foam-type, which also was new.

Holley Tests
October 1971

It also increases air-inlet noise.

Note that some air cleaners will allow full airflow capability. These are the air cleaners that should be used by racers, even if their use requires adding a hood bulge. An air cleaner that gives full-flow capability to the carburetor also gives impressive high-rpm power improvements, as compared with one that restricts flow.

It is important to check the clearance between the upper lid of the air cleaner and the top of the carburetor's pitot or vent tubes. The height of air cleaner elements varies as much as 1/8 in. because of production tolerances. The shorter elements can place the lid too close to the pitot tubes so that correct bowl reference pressures are not developed. Whether the pitot tubes are angled or flat on top, there should always be at least 3/8 in. minimum clearance between the tip of the pitot tube and the underside of the air cleaner lid.

You have noticed the long *snorkel* intakes on modern air cleaners. These are designed to reduce intake noise—not to improve performance. High-HP engines usually have two snorkels for more air and perhaps for "more image," too. For competition, the snorkel can be removed where it joins the cleaner housing. Additional holes can be cut into the cleaner housing to approximate an open-element configuration to improve breathing. Or, it is sometimes possible to expose more of the element surface by inverting the cleaner top.

When it comes time to race, use a clean filter element. Keep the air cleaner base on the carburetor if you possibly can, even if you have removed the air cleaner cover and element. Be sure to secure the base so that it cannot vibrate off to strike the fan, radiator or distributor.

Because the carburetor is internally balanced, that is, the vents are located in the air-horn area, no jet change is usually required when the air cleaner is removed.

VELOCITY STACKS

Velocity stacks are often seen on racing engines. These can improve cylinder filling (charging) somewhat, although this depends on many other factors. For instance, when velocity stacks are used on carburetors on an isolated-runner manifold, the stacks may form part of a tuned length for the air column.

Velocity stacks also have a straightening effect on the entering air. And, they can

Two stacked air cleaners with fresh air plumbed from forward of radiator. Good package for any hard-working engine.

Additional ducting (arrow) can cause bowl-pressure problems. Original snorkel is plugged and air filtered through remote cleaner and stock element in this Baja 1000 mile race installation.

Large, fully open air cleaner used on high performance Buick V6 equipped with Q-jet.

contain fuel standoff, which can occur with isolated runner manifolds. Remember that velocity stacks need space above them to allow air to enter smoothly. Mounting a hood too close to the top of the stacks will reduce airflow into the carburetor. Provide a minimum clearance of 2 in. between the top of a velocity stack and any structure over it.

When an air cleaner is used on a carburetor equipped with stacks, keep the 2-in. clearance between the top of the stacks and the underside of the air cleaner lid. This may require using two open-element air cleaners fastened together with sealant—and a longer stud between the carburetor top and the cleaner lid.

Whevever you have seemingly insurmountable metering problems at high rpm and WOT, keep in mind that they might be caused by the air cleaner and not metering.

Velocity stacks on four Rochester two barrels used for racing. No air cleaners here.

DISTRIBUTION

When a manifold or carburetor change is made, check distribution to determine if all cylinders are receiving an equal mixture. Even though a previous carburetor and manifold combination may have supplied nearly perfect distribution, you cannot take the chance that one or more cylinders will be running lean.

Although there are several test methods to check distribution, you have limited resources when at a race. You must rely on the appearance and color of the plug electrode and porcelain base, and you can also observe the color of the piston tops with an inspection light. These examinations give valuable information about combustion and A/F distribution in the engine. To accurately inspect the plugs, buy an illuminated magnifier so you can see the base of the porcelain.

Checking plug color gives a rough idea of what is occurring with mixture ratio and distribution. It is meaningful for WOT performance only if checking is done after driving at WOT, and stopping the engine while simultaneously closing the throttle.

Using plug color to check distribution will only be helpful when the engine is in good condition. Engine condition can be checked with a compression test to make sure that all cylinders have equal compression at cranking speed. Or, for a more accurate check, a leak-down test can be used to compare cylinder condition. Gages to do

these tests are available from many tool companies.

Remember that new plugs take time to acquire color, and as many as three or four drag-strip runs may be needed to do so. Plug color is only meaningful when the engine is declutched and shut off at the end of a high-speed, full-throttle, high-gear run. If you allow the engine to slow while still running, plug appearance will be meaningless.

Plug readings can be made after full-throttle runs on a chassis dynamometer (dyno), with the transmission in an intermediate gear, but road tests require high gear to correctly load the engine. Similarly, plug checks can be made where the engine has been running at full throttle against full load applied by an engine dyno.

It is much easier to get good plug readings on the dyno because full power can be applied and the engine shut off without difficulty. Plugs can be read quickly because you can get to them easier than in the usual car or boat installation. However,

READING SPARK PLUGS
Rich—Sooty or wet plug bases, dark exhaust valves.
Correct—Light-brown (tan) color on porcelains; exhaust valves are a red-brown clay color. Plug bases are slightly sooty—base leaves a slight soot mark when turned against palm of hand. New plugs start to color at base of porcelain inside shell: This can only be seen with an illuminated magnifying viewer. For drag engines with wedge (quench-type) combustion chambers, speed and elapsed time become more important than the plug color. Plugs may remain bone-white with best speeds/times. A fuel mixture that gives the light-brown color may be too rich for these type of engines.
Lean—Plug base is ash-grey. Glazed-brown appearance of porcelains may also indicate plug's heat range is too hot. Exhaust valves are a whitish color.
NOTE: Piston-top color seen with an inspection light through the plug hole can be a quicker—and sometimes more positive—indicator of mixture, than plug appearance. Careful tuners look at *all* indicators to take advantage of every possible clue to how the engine is working.

113

don't think that plug heat range and carburetor jetting established on an engine dyno will be absolutely right for the same engine installed in your racing boat or chassis. Airflow conditions past the carburetor can easily change the requirements.

It would be nice if every plug removed from an engine looked like the others from the same engine, in color and condition, but this is seldom ever achieved! Color and other differences indicate that combustion-chamber temperatures and A/F ratios are not the same in every cylinder.

If differences exist in the firing end of the plugs when you examine them, the cause may be due to one or more factors: unequal distribution of the mixture, unequal valve timing due to incorrect lash or a worn cam, or poor oil control caused by rings or excess clearance.

Problems within the ignition system that can lead to plugs not reading the same or misfiring include: a loose point plate, arcing in the distributor cap, and defective rotor cap, plug wires or connectors. Firing between plug wires, a defective primary wire, or even a resistor that opens intermittently also cause problems.

Pay special attention to the cleanliness of the entire ignition system, including the inside and outside of the distributor cap and the outside of the coil tower. Also, clean the inside of the coil and cap cable receptacles. Any dirt or grease here can allow some or all of the spark energy to leak.

If the cylinders have equal compression and the valves are lashed correctly, a difference in plug appearance may indicate that there is a mixture-distribution problem. It is sometimes possible to remedy this with main-jet changes.

For instance, if one or more plugs show a lean condition, install larger main jets in the throttle bore(s) feeding those cylinders. Should all of the plugs appear to show a rich condition, install smaller main jets in the throttle bore(s) feeding those cylinders.

A difficult problem occurs when cylinders fed from the same plenum show different mixture conditions. Some are lean and some rich, or perhaps a combination of these conditions. This indicates a manifold fault that cannot be corrected with jet changes. Correcting such conditions requires manifold modifications that are beyond the scope of this book.

COLD AIR & DENSITY

Higher-density inlet air improves the engine's power capability in proportion to the density increase. Let's examine the practical aspects. What can you do to keep the density "up" to get the best HP from your engine?

First, the underhood temperature is not ideal for HP production because, even on a reasonably cool day, air reaching the carburetor inlet has been warmed by passing through the radiator and over the hot engine components. Underhood temperatures soar

to 175F (80C) or higher when the engine is turned off and the car is standing in the sun. An engine ingesting warm air will be down on power by more than you might imagine.

Assume that the outside (ambient) air temperature is 70F (21C) and the underhood temperature is 150F (66C). Use the following equation:

$$\gamma_{oa} = \frac{460 + t_{uh}}{460 + t_{oa}} \times \gamma_{uha}$$

$$\gamma_{oa} = \frac{460 + 150}{460 + 70} \times \gamma_{uha}$$

$$\gamma_{oa} = 1.15\ \gamma_{uha}$$

where

γ_{oa} = outside air density
γ_{uha} = under-hood air density
γ_{oa} = outside air temperature
γ_{uh} = under-hood air temperature

In the example here, outside-air density is 115% of the underhood-air density, or 15% greater. Available horsepower will increase by an equivalent amount. If the engine produces 300 HP with 150F (66C) air-inlet temperature, it can be expected to produce 345 HP with 70F (21C) air-inlet temperature, provided the carburetor meters the correct amount of fuel.

Cold air gives more improvement than ram air because about 1% HP increase is gained for each 5F decrease in temperature. This assumes the mixture is adjusted to compensate for the density change and that there is no detonation or other problems.

If air scoops are being used to duct air to the carburetor, do not connect the hose or scoop directly to the carburetor. Instead, connect the scoop to a cold-air box or to the air cleaner housing to avoid creating turbulence in the incoming air as it enters the carburetor air horn. Be sure to retain the air cleaner element as a diffuser to also reduce the incoming air's turbulence. Otherwise, a high-speed miss may occur.

If you plan to build an intake duct that picks up cold air at the front of the car, be prepared to change the air cleaner filter element on a regular basis, maybe as often as once a week in dustier areas. If you eliminate the filter altogether, plan on new rings or a rebore job soon, because the engine will quickly wear out.

Duct cold air from the cowl just ahead of the windshield. This is a high-pressure zone that will supply cool outside air to the carburetor. This area still gets airborne dust, but it is several feet away from the dirt and grit at road level.

Massive fresh-air hood draws air from high-pressure zone at base of windshield to Q-jet on Buick V6.

Hoods with intake scoops can be mated with a tray under the carburetor(s) to ensure that warm underhood air cannot enter the carburetors to reduce inlet-air density. As mentioned earlier, consider using air cleaners as diffusers.

Drag racers have done plenty of experimentation on scoops and fresh-air supplies from the cowl area ahead of the windshield. One popular modification has been to reduce the entry area by gradually diminishing the cross-sectional area of the scoop or air box from the plenum. This increases the velocity of the air being supplied to the carburetors, but it may also increase air turbulence. It then becomes harder to turn the air smoothly to enter the carburetor.

Drag racers often keep the hood open and avoid running the engine between events so that the engine compartment stays as cool as possible. It is helpful to spray water onto the radiator to help cool it. This ensures that the engine water temperature will be usable and that the radiator will not heat incoming air any more than is absolutely necessary.

For most racing, the engine coolant temperature should be around 180F (82C). This level helps reduce friction inside the engine. The ideal is to keep engine oil and water temperatures at operating levels while taking care to keep down the inlet-air temperature.

A heated manifold reduces density. Exhaust heat or jacket water to the manifold should be blocked off to create a cooler manifold when performance is being sought. There are various ways of accomplishing this.

In some instances, there will be intake-manifold gaskets available that close off the heat openings. Or, block the opening by slipping a piece of stainless steel or other metal between the gasket and manifold. Many competition manifolds have no heat riser so the manifolds are cooler to start with.

Intake manifold being cooled between races with wet towel to improve density and increase power.

Some car makers offer shields that fit under the manifold to prevent heating the oil with the exhaust-heat crossover passage. These shields also prevent the hot oil from heating a cold manifold, or at least reduce that tendency. This can be a cheap way to gain approximately 10 HP on a small-block V8 engine.

If the manifold is water-heated, it is a simple matter to reroute the hoses so the water does not flow through the manifold.

RAM TUNING

Ram tuning can give better cylinder filling, and thus improved VE, in a narrow speed range by taking advantage of a combination of engine-design features:
- Intake- and exhaust-system passage or pipe lengths.
- Valve timing.
- Velocity of intake and exhaust gases.

Although ram tuning improves torque at one point or across a narrow rpm band, the improvement tends to be "peaky," so that the power falls off sharply on either side of the peak. It is generally understood than ram tuning is based on *resonance*. Reso-

nance is sought at a tuned peak, with the knowledge that *the power gained at that point will be offset by a corresponding loss at other speeds.*

Individual intake-manifold passages for each cylinder (ram or tuned length) can be measured from the intake valve to a carburetor inlet if there is one venturi per cylinder. Or the length may be measured from the intake valve to the entry to a plenum chamber fed by one or more carburetors.

The longer the passage length, the lower the rpm at which peak torque will occur. Making the manifold-passage size, intake ports and valves larger *raises* the rpm at which best filling occurs. Best driveability and street performance are usually achieved with small-port manifolds and heads.

HEADER EFFECTS VS. JETTING

Headers will usually reduce exhaust back pressure so that the engine's volumetric efficiency is increased (it breathes easier). The main effects of headers will be seen at WOT and high rpm. In most instances, no jet changes will be needed if the carburetor has been correctly

---

INCREASED DENSITY MEANS LARGER JETS

No matter how a density increase is obtained, the increase must be accompanied by the use of larger main jets. The size increase (in area) is directly proportional to the square root of the density increase (in percent).

sized for the engine displacement and rpm. The carburetor will automatically compensate for any increased airflow by increasing the fuel flow in the correct proportion. The mixture ratio will not be affected.

Q-JET TUNING FOR PERFORMANCE

Calibration changes and modifications discussed here are intended for *off-road use only* and generally apply to carburetors that are not microprocessor-controlled (pre-'81). The newer electronic carburetors don't lend themselves to many high-performance modifications; you are better off leaving them stock in most cases. They achieve acceptable performance, good economy and limit emissions in their original form.

Before going into the specific Q-jet modifications, I'll discuss some other variables that influence their performance. It is important that you understand what is taking place overall and how these affect your efforts in changing the internals of a Q-jet.

Don't get eager to rip into the carburetor and turn to this book's modifications section with tools in hand. Study these next few pages and relate the discussion to your application. There aren't enough pages to cover all applications, but these recommendations hold for most.

FUEL PRESSURE

The first order of preparation is to be sure adequate fuel gets to the carburetor bowl at all times. The preferred fuel pressure is 4—6 psi. The Q-jet can function well under certain conditions, such as over rough terrain and at part-throttle, with as little as 2 psi. The ideal setup is to have a fuel pump

and lines of adequate capacity so the fuel pressure never drops below 4 psi or exceeds 6 psi.

In some high-performance applications, there may be times when fuel pressure needs to be 8—10 psi. When pressures above 6 psi are used, flooding can occur during idle and slow-speed driving.

Never design a fuel system for a street-driven car with more than 6-psi fuel pressure at idle because the fuel level rises with increased pressure. This causes idle instability, over-rich low-end mixtures and hot-start problems.

Keep in mind when evaluating fuel pressure that the pressure measurement is being taken at a point outside (ahead of) the fuel bowl. If fuel demand through the jets becomes high enough and the fuel inlet valve is too small to supply the need, you will be led down the path of poor performance, with plenty of pressure showing on the gage. Correctly sizing the fuel inlet is one of the most important parts of carburetor performance tuning.

Use A Fuel Pressure Gage—It's best to monitor fuel pressure on a 0—10 psi gage mounted outside the cockpit of the vehicle. A 0—30 psi gage of good quality is also acceptable for high-pressure applications. Fuel lines inside the vehicle can be cut or ruptured during an accident or by mechanical failure. If the gage must be placed inside the cockpit, use an electric one. The gage should be connected to the fuel line as close to the carburetor inlet as possible.

Low fuel pressure can be caused by:
- Fuel pump with insufficient capacity.
- Fuel line too small.
- Restrictive fuel line fittings.
- Too many bends or a crimped line.
- Tank outlet or filter screen restricted.
- Plugged fuel filter.
- Filter installed between the tank and the fuel pump.
- Dirt anywhere in the system.
- Lines too close to a heat source.
- Incorrectly vented cap on fuel tank.

MODEL B, 2G, 4G & H MAIN METERING JETS

Three types or groups of main metering jets are used on these models to obtain accurate calibration. Any replacement jet *must be of the same group* as the original equipment jet to avoid changing the flow calibration. The group in which each jet belongs can be readily identified by examining the approach angle (nearest to surface with stamped jet number). For example, it is possible to have a jet from each of these groups A, B, C with the number 60 stamped on the jet face: A—7002970; B-7002660; C—7008660.

Group A Main Jets
90° Square Approach (Zinc Plate)

Part	Stamped On Jet	Part	Stamped On Jet
7002943	43	7002956	56
7002945	45	7002957	57
7002946	46	7002958	58
7002947	47	7002959	59
7002948	48	7002960	60
7002949	49	7002961	61
7002950	50	7002963	63
7002951	51	7002964	64
7002952	52	7002965	65
7002953	53	7002966	66
7002954	54	7002967	67
7002955	55	7002969	69

Group B Main Jets
60° Approach (Short Taper)

Part	Stamped On Jet	Part	Stamped On Jet
7002639	39	7002651	51
7002640	40	7002652	52
7002641	41	7002653	53
7002642	42	7002654	54
7002643	43	7002655	55
7002644	44	7002656	56
7002645	45	7002657	57
7002646	46	7002658	58
7002647	47	7002659	59
7002648	48	7002660	60
7002649	49	7028554*	54
7002650	50	7028559*	59

Group C Main Jets
60° Approach (Long Taper)

Part	Stamped On Jet	Part	Stamped On Jet
7008660	60	7008679	79
7008661	61	7008680	80
7008662	62	7008681	81
7008663	63	7008682	82
7008664	64	7008683	83
7008665	65	7008684	84
7008666	66	7008685	85
7008667	67	7008686	86
7008668	68	7008687	87
7008669	69	7008688	88
7008670	70	7028660*	60
7008671	71	7028661*	61
7008672	72	7028662*	62
7008673	73	7028663*	63
7008674	74	7028665*	65
7008675	75	7028669*	69
7008676	76	7028670*	70
7008677	77	7028673*	73
7008678	78	7028686*	86

*Stainless Steel

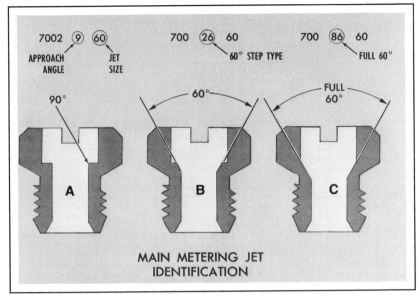

Examine *approach angle* of Model B, 2G, 4G & H main jets to identify group. **Replace jets** *only* with those from same group.

MONOJET MAIN METERING JETS

The Monojet main jet is different from other RPD models and should not be interchanged. It has five radial lines stamped opposite the jet identification number. The number indicates orifice size; it can be two or three digits, depending on orifice size. Jet size can be determined by subtracting 100 from the last three digits of the part number. For example:

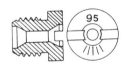

Part	Stamped On Jet	Diameter of Orifice
7034195	95	.095"

MONOJET MAIN METERING JETS

Part	Stamped On Jet	Part	Stamped On Jet
7034154	54	7034203	103
7034192	92	7034204	104
7034195	95	7034205	105
7034198	98	7034206	106
7034199	99	7034215	115
7034200	100	7034218	118
7034201	101	7034225	125
7034202	102	7034228	118

MONOJET MAIN METERING RODS

The Monojet main metering rod diameter is a three-digit number stamped on the shank of the metering rod. Example:

Tip diameter is 0.048-inch.

Part	Stamped On Rod	Part	Stamped On Rod
			Stamped 060, Diameter A .060"
7035929	070	7037662	066
7035930	072	7037974	076
7035931	074	7037979	094
7035932	080	7040630	078
7035934	088	7040770	060
7035935	096	7044791	081
7036226	086	7045843	098

Fuel Lines—These must be routed away from heat and firmly secured so that they will not vibrate and fatigue. Fittings must be nonrestrictive. Some fittings have internal restrictions, so check every fitting in the fuel supply system. Make sure that the passage is the same size all the way through the fitting. If it tapers inside, you may be able to open the passage with a drill.

Avoid 90° (right-angle) fittings wherever possible because these restrict fuel more than straight or 45° fittings. A right-angle fitting has the same restrictive characteristics as a piece of tubing several feet long.

If a mechanical fuel pump is used on a competition, large-displacement engine, install a 1/2-in. inside-diameter (ID) line from the tank to the fuel pump. Make sure the fitting connecting to the fuel tank is as close to the fuel-line ID as is consistent with safety.

Attaching the correct-size fuel line to the fuel tank is one point commonly overlooked by mechanics just getting started in competition. They install a large line, but try to feed fuel through a tiny tank opening—1/4 in. or smaller in diameter. This is a poor practice.

Here is a point where steel, copper or aluminum tubing is a better choice than fuel hose because tubing fittings allow a larger passage than hoses. This is because a hose fitting has to have enough wall thickness to withstand the pressure of hose clamps. Fittings for tubing can have openings that are very close to the actual ID of the tubing itself, thus giving less restriction.

COPPER, ALUMINUM OR STEEL FUEL LINE

Copper and aluminum tubing will fatigue quicker than steel. Use steel line for installations you want to last. If copper or aluminum is used, mount it so flexing is limited.

Steel fuel line with no flexible section is what you'll find on GM cars and trucks in nearly all instances. This is the recommendation from RPD engineers because they believe flexible line is more subject to failure. But metal fuel lines can fail due to kinking, perforation by rust, or vibration fatigue.

If you install a section of fuel hose or other flexible line, remember to check it and its clamps regularly. Brittle and cracked fuel hoses are a fire hazard anywhere in the fuel supply system.

Adjustable fuel pressure regulator between fuel pump and carburetor ensures against excess pressure that causes hot-starting and flooding problems.

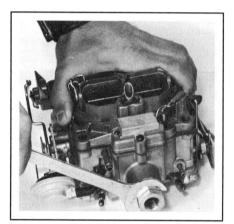

MAC Tools S-141 is slim wrench ideal for turning fuel inlet nuts.

Mount mechanical fuel pressure gage outside driver's compartment. Not pretty, but much safer than having ruptured line spewing fuel inside.

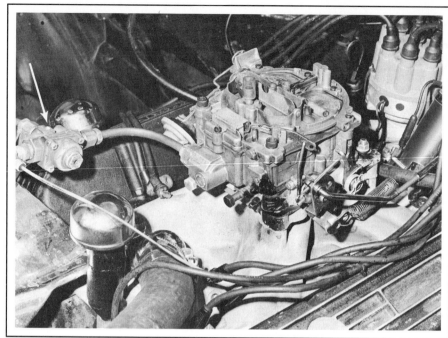

Holley fuel pressure regulator (arrow) used with modified Q-jet adapted to Edelbrock manifold.

Drag racing engine using steel braided fuel line from twin Holley fuel pressure regulators to fuel inlet.

The important point is to keep the fuel supply to the pump unrestricted. The pump outlet for an engine-mounted mechanical pump or the regular outlet connection to the carburetor(s) can be through 5/16- or 3/8-in. tubing. The latter is preferable.

Keep all fuel lines away from exhaust components to avoid excessive heat. Make sure no part of the vehicle body will deflect exhaust from open headers back into an area where the line is mounted. If a line passes near the exhaust system and there is no other place to route it, thoroughly insulate the fuel line. It should be clamped against the chassis or body structure with rubber-lined clamps such as those used in aircraft repair shops.

Locate the pressure regulator—if one is used—as close to the carburetor as possible. Lines between the regulator and carburetor can be 3/8-in. ID. There are usually two outlets on regulators designed for high performance. Where two carburetors are used, connect each carburetor to an outlet.

Pressure at the carburetor must be set with the engine idling so that there will be some fuel flow to allow the regulator to function. Use a fuel pressure gage at the carburetor and adjust the regulator to supply the desired pressure.

Fuel in the pump may become heated when it is pumped from the outlet, through a bypass and back into the pump inlet. To avoid this problem, which is most apparent at idle, some racers install an external bypass through a 0.020—0.060-in. restriction from the pump outlet to the tank. Don't use a bypass restrictor any larger than needed or you could reduce the pump output too much.

The bypass line should dump back into the main tank low enough to enter liquid fuel. If the bypass fuel is dumped in above the fuel level it picks up air and creates vapor. The bypass can also be used with mechanical pump installations. In this case, the small-diameter return line is usually connected from the carb inlet to the fuel tank. Some AC fuel filters are fitted with a bypass-line connection that is handy for this plumbing.

Fuel Filters—A fuel filter should be used between the fuel pump and the carburetor. Never use any kind of filter, other than a simple screen, on the suction side of a fuel pump. This is true regardless of the type of fuel pump. The usual screen at the tank outlet will work fine if it is clean.

The filter in the line to the carburetor should be as nonrestrictive as possible. Paper-element filters are excellent for this purpose. If concerned about the pressure drop through the filter, "Y" the fuel line to run through two filters so each one supplies a carburetor, or use two filters in parallel in the line to a single carburetor. Mount the filter canister to allow easy replacement of the element. Large truck and race-designed filters are available and recommended for performance applications.

Sintered-bronze (Morraine) filters found in the inlets of many carburetors are fine for street use *if* there is little dirt in the gasoline and *if* the tank itself is clean. If the fuel supplied to the pump is dirty, as may be the case in a dusty area, remove the sintered-bronze filters from the inlets and install an in-line filter between the pump and the carburetor.

Fuel Pumps—There are two common types of fuel pumps: mechanical and electrical. Although the carburetor never "knows" which kind supplies its fuel, let's examine their respective merits and debits.

Mechanical fuel pumps are engine-mounted pumps with internal diaphragms driven by camshaft/crankshaft eccentrics operating a rod or lever. Advantages include low initial cost, simple plumbing and mounting, low noise level, and familiarity to the general public because millions have been used over the years. The primary disadvantage is the transferal of engine heat

Remove filter elements from fuel inlet for competition efforts.

Put fuel pressure monitoring point as close to inlet needle as possible. Q-jet lends itself to clean installation (arrow). Thin casting allows only a few tapping threads, so epoxy fitting to strengthen.

Install in-line filter just ahead of fuel inlet.

Cutaway shows route (arrows) fuel takes through stock carburetor inlet filter. Performance applications require less restriction so remove sintered-bronze filter.

into the fuel, especially when the engine is stopped hot. The mechanical pump sucks fuel through a long line from the tank, further promoting the fuel's tendency to flash into vapor, especially on warm days. Fuel vaporizes more readily with decreased pressures.

Electrical fuel pumps were not widely used as OEM equipment until the early '70s. Electric pumps in the fuel tank become more popular each year. An externally or frame-mounted electrical pump has a high initial cost, is noisier than a mechanical pump—unless special mounting procedures are used—and requires connection to the car's electrical system. It should be mounted near the tank, away from heat and flying rocks, and plumbed to the tank outlet and to the fuel line.

Electrical pumps are typically mounted at the rear, near the tank, so they push the fuel. Fuel pushed forward to the carburetor has less tendency to flash into vapor as it moves through the line. For competition or hot-weather use, the rear-mounted electrical pump helps prevent vapor lock. There are several types of electrical fuel pumps: solenoid-operated (AC, Bendix, Lucas), vane-type (AC, Holley) and positive-displacement (gear-type).

When installing an electrical pump, eliminate the mechanical pump, if at all possible. It heats fuel and limits total fuel-supply pressure to the mechanical pump's output pressure.

Include a safety switch in the circuit so the pump will not work unless there is oil pressure. For starting the engine before oil pressure develops, run a wire so the starter energizes the pump from the starter-solenoid circuit. Once the engine is running, the three-way switch sends voltage to the pump as long as there is oil pressure. A schematic showing the wiring from the switch is included here for reference.

Mount the pump in as cool an area as possible: keep it away from any exhaust components. Any fuel pump mounted to the chassis will transmit some noise into the car's body structure. This is no problem on a race car, but it can be a source of annoyance on a street car.

The best way to reduce the transmission of pump noise into the body is to mount the pump on rubber-insulated studs. These are mounted in turn on a chassis member or on a stiffened section of the body. Never mount the pump directly to a large flat

This AC mechanical fuel pump can be rebuilt because valve and diaphragm housing is screwed to pump casting. Often used on competition cars when rules require mechanical pump.

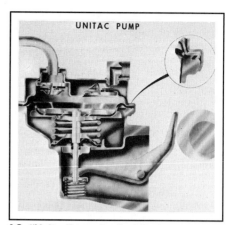
AC "Unitac" mechanical fuel pump isn't rebuildable because valve and diaphragm housing is crimped to pump casting.

AC EP-11/12 electric fuel pumps supply up to 22 gallons per hour at 4.5—8 psi fuel pressure. Maximum current draw is 3.5 amperes.

AC EP-1/2 electric fuel pumps supply up to 35 gallons per hour at 4.5—8 psi fuel pressure. Maximum current draw is 2.0 amperes.

AC PS-9 oil pressure safety switch for electric fuel pumps.

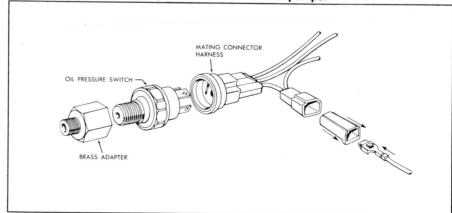

sheet-metal surface because this will amplify the noise, just the opposite of what you are trying to achieve.

There's a good reason to mount the pump near the tank. A pump trying to pull fuel creates a vacuum on the end of a fuel line (especially a long one), which tends to cause the fuel to flash into vapor. This is why professional racers replace the engine-mounted mechanical pump with an electric pump near the fuel tank.

When fuel is pumped forward at high pressure, there is another advantage that may not be quite so obvious. The high pressure more than offsets pressure losses in the lines from friction and acceleration. Pumping the fuel forward at high pressure and then reducing its pressure before sending it to the carburetor, ensures that there will always be adequate pressure available at the carburetor.

I'll emphasize one point: Keep the fuel pump inlet line short and large. If a filter is used between the fuel tank and the pump inlet, it must be a screen filter so there is no pressure drop at this point, which could cause fuel vaporization and pump cavitation. It is useful to place the pump so its inlet is slightly below the tank so that the fuel level will create a positive pressure at the inlet. On a drag car, it helps to place the pump behind the tank so acceleration will tend to push fuel to the pump inlet.

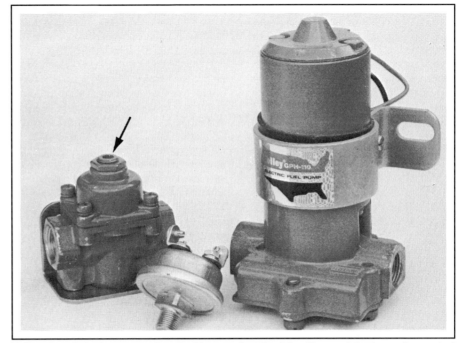

Holley electric fuel pump pushes fuel from tank at 9—14 psi to regulator (arrow) mounted near carburetor. Pressure switch in center (Holley 89R-641A) senses engine oil pressure to shut off fuel pump when engine is stopped, regardless of whether ignition is turned off or not.

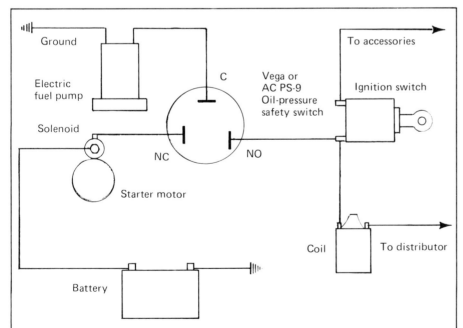

Typical wiring diagram for electric fuel pump. AC oil pressure switch (Chevrolet 3986857) stops pump if oil pressure drops. This wiring starts pump when ignition switch energizes starter solenoid. Starter circuit from ignition switch is not shown.

Example of OEM fuel supply design with electric fuel pump (circled) mounted in gas tank. Pump weighs five oz., has 1-3/8-in. diameter and 3-3/4-in. length.

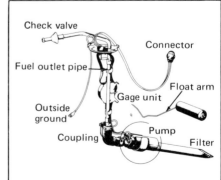

FUEL BOWL

The Q-jet fuel bowl has adequate capacity for general driving and most performance uses. It is shaped well and located for good cornering and rough-terrain stability. Under certain conditions it does need improvements to keep fuel from spewing from vents.

Fuel Spewing—This spewing generally occurs during WOT accelerations at about 3500 rpm. You will notice a sag (loss of power) that sometimes continues for a number of seconds.

It can be caused by two conditions. Certain air-cleaner or inlet-air plumbing configurations can cause air velocities to act as a depression on a vent area and allow it to feed fuel. This is uncontrolled metering you cannot tolerate. Second, rapid fuel entry through the inlet valve can cause pressure spewing.

When a depression (vacuum) exists, as in the first case, the vents are able to grasp fuel that is being sprayed wildly inside the fuel bowl from the inlet needle. This fuel is then expelled upward from the vents, creating a rich mixture.

Vent-spewing problems can be eliminated in many cases by extending the vertical vent tubes or cutting the top of the tube at a 45° angle.

Try cutting the tube first; it's easiest and may fix the problem. This causes air to pack the vent stack with minor pressure, helping to keep fuel in the bowl. Should it fail to eliminate spewing, increase the length of the tube up to 1/2 in. You can also add a vent stack to the rearward internal vent if need be.

If intake air is routed in from the back side of the carburetor, the angle-cut may have to be open to the rear. Some ex-

perimentation may be needed to keep a positive pressure on each of the vent stacks.

Sometimes special engine adaptations create a need for fabricated air cleaners and unique ducting of air to the carburetor. This can cause detrimental air-ram or depression effects on carburetor venting. Some serious thinking and "baffle engineering" (guess and try) followed by component fabrications can eliminate the problem.

Loss of power (sag) at WOT or heavy acceleration after hot soak can sometimes be eliminated by installing deflector (note pencil tip).

Fuel flow decreases as lift (supply to inlet) and head (pump to carburetor) distances increase. Maximum flow is at small head and lift measurements.

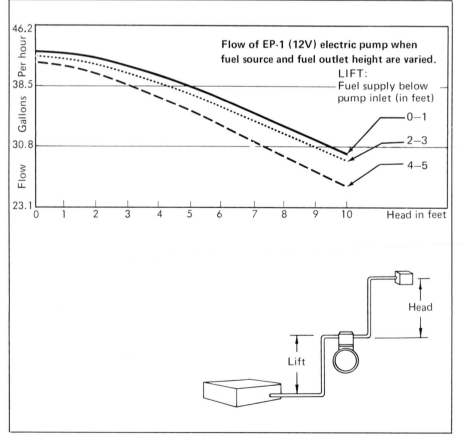

Flow of EP-1 (12V) electric pump when fuel source and fuel outlet height are varied.

LIFT:
Fuel supply below pump inlet (in feet)

0–1
2–3
4–5

Head in feet

A few Q-jets have 45° slashed internal bowl vent. Vent spewing can be curtailed by increasing its height and adding tube in existing rear vent hole (arrow).

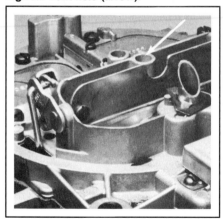

Vent spewing can sometimes be prevented by installing a 100-mesh screen over the vents. The screen helps separate the air and fuel, causing the fuel to drop to the bottom of the float chamber more directly than if allowed to wash over the air-horn gasket. The screen can also be effective in reducing or eliminating vapor lock because it condenses fuel from highly vaporous gasoline entering the fuel bowl.

Last, but not least, is the plastic fuel stuffer. If you left it out during initial preparation, keep it handy. Sometimes it is all you need to direct incoming fuel downward to the metering orifices. I never recommend leaving out the deflectors mounted around the inlet needle.

Hot Fuel Flash—Another problem usually results from hot soaking during brief stops. It is a fuel-flash condition. If a car is run at high speeds in hot weather then stopped for 5 to 20 minutes, restarted and accelerated at WOT back to these speeds, it may falter and die out momentarily if the fuel is high in vapor.

This occurs when the engine exhaust manifolds, combustion chamber and other large pieces get very hot during continuous running. This condition multiplies rapidly in a race car as continuous high revs create additional heat. When the vehicle is stopped for this brief period, the heat from these large iron areas is transmitted to water and engine-dress accessories that remain cooler during running because of airflow and radiator cooling.

Gasoline boils at approximately 150F (66C). Consequently, when fuel pump and carburetor temperatures rise to approximately 190F (88C), the fuel vaporizes and escapes from the bowl. After you restart and maneuver from rest, the bowl fuel is replenished with hot unstable fuel from the pump and lines.

The float, vents and metering can usually handle this heat as long as you do not go immediately to WOT. At best, you may have to cope with some stalling and an unstable idle. This annoying problem is most likely to arise if you do accelerate to WOT within a minute or so of a restart.

Cooler fuel is brought in from the fuel tank and pressurized ahead of the fuel-inlet needle. As you accelerate hard, the fuel in the bowl needs to be replenished rapidly. The float drops and a rush of fuel enters. When this fuel enters the hot bowl suddenly, it can flash to vapor so violently it will spew from every outlet. The engine cannot cope with this additional unmetered mixture and will sag or die.

When a vehicle is acting up after a hot soak, it's best to run it a couple of minutes at a fast idle in neutral before putting it in gear. Give the carburetor and other accessories a chance to transfer some of their heat into the cooling water or air stream.

The term *fuel flash* doesn't refer to flame or imply fire hazard. As long as the air cleaner and original equipment pieces are intact, these vaporous fuel flashes are arrested. Should you be foolish enough to be running without an air cleaner, with altered pieces or inadequate bolt-on accessories, a fire could well be the result. Emitting this flammable vapor underhood invites it.

In summarizing problems in the fuel bowl, note that a rich fuel flash or a lean vapor lock can cause the engine to lose power or die following a hot soak. Separating the two can sometimes be frustrating.

The best diagnostic approach may be to

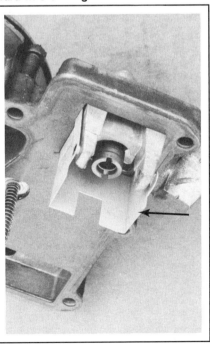

Tin shield (arrow) is effective in forcing inlet fuel into bowl rather than outward and upward where it might reach internal vents.

get a friend to follow you through an exercise that produces the problem. If it's fuel-flash richness, a large puff of black smoke will emit from the exhaust during disrupted engine operation. If it's lean vapor lock, smoke is unlikely. Heavy blue smoke doesn't count because that is caused by oil burning.

FUEL INLET VALVE

For most RPD carburetors there are various sizes of inlet valves available. In the case of the Q-jet, Delco supplies three sizes. Following is a parts listing for these units:

RPD	DELCO	ORIFICE SIZE
7035130	30-130	0.110-in.
7035140	30-140	0.125-in.
7035142	30-142	0.135-in.

If your Delco dealer is not able to supply you with suitable parts, write:

Roe, Inc.
P.O. Box 26848
Tempe, AZ 85282

These carburetor specialists have inlet valves up to 0.140 in. at approximately 0.005-in. increments. Don't use off-brand parts. Be particularly skeptical of newly designed items for which revolutionary improvement claims are made. Most such items are products of good marketing, not good design.

Don't install an inlet valve that is larger than necessary. Flooding is the most common result. This causes driving misery in the pits or on the grid at drags or road races, and guarantees untold grief for off-road enthusiasts.

The second problem that happens more often than most people are aware of is internal-vent spewing.

Fuel behaves erratically when it enters the bowl area too rapidly. As the fuel passes the inlet-valve seat, it is leaving a pressure area of 4—8 psi and entering a bowl area at near atmospheric pressure (zero).

Unless it is cooled adequately with a fuel cooler or is of low volatility, the fuel will likely turn to vapor. It can travel along the top of the bowl-gasket surface and actually feed into the air stream via the internal vent stacks. Remember too, some intake-air fabrications can cause depressions at the internal vents. This is quite common during heavy-throttle operation and is seldom diagnosed correctly.

The flat/dead feel of the engine under these conditions is often called *lean-out*. I have made vehicles with this problem go quicker by simply reducing the inlet-valve size. Be aware of this potential fuel spewing from the vents and try an easy fix first: a smaller inlet valve.

The carburetor's capability to control fuel level correctly under most conditions is influenced by these primary factors:

- Fuel-bowl size and shape.
- Buoyancy and shape of the float.
- Geometry of the float arm to the inlet needle.
- Inlet-valve seat-orifice size.
- Fuel pressure at the inlet valve.

The first three items are established in the original design. Once the bowl is designed, major changes in float size are limited. Buoyancy changes can be accomplished to some degree by introducing new materials, new shapes, or adding assist springs.

The following inlet-valve/float-level settings are recommended for all-out competition:

Float	Float Setting*	Inlet Valve
7034454 long fulcrum	0.325-in. at 6 psi fuel pressure	30-140 (0.125-in.)
7037316 short fulcrum	0.296-in. at 6 psi fuel pressure	30-134 or 30-142 (0.135-in.)

*Higher fuel pressure may require increasing these settings to maintain the same fuel level in bowl.

I have known sharp oval-track mechanics who intentionally design the entire fuel system to lean the A/F mixture toward the end of each straightaway. They know a slightly lean power mixture (14:1 or 15:1) gets maximum performance.

For continuous power demands it would not be wise to calibrate lean because it causes high combustion chamber heat. A mechanic who sets up with a good, safe power mixture (rich side) for most of the course, with an automatic leaning out at the end of each straightaway, gets a double bonus:

- More power for a few seconds on the straight.
- Better economy (less fuel consumption).

This can mean winning or losing if you are limited on fuel capacity by the race-association rules.

Should you choose to lean out the mixture intentionally for short-spurt performance, it is best to do so by altering the fuel inlet valve size. Gradually decrease the size until you note a slight loss in performance. Monitor this carefully with a stopwatch, on a dynamometer, or during practice at the race track.

If you attempt to do this size evaluation on an engine dynamometer, I suggest you rig a sight glass in the bowl so you can see if the smaller valve maintains suitable levels. Remember, dyno steady-state running is different from acceleration conditions that are dynamic (constantly changing). There is no better way to monitor performance than under actual driving conditions.

Once you get a slight drop in performance, come back up in inlet size a few thousandths to reach the ultimate point of performance for the driving application.

As mentioned before, you can go small enough in inlet valve size to where metering demand will use fuel from the bowl faster than it's replenished. If there is too much restriction between the pressure-gage pickup and the bowl (inlet valve), the gage will be reading good pressure while the nozzles (fuel bowl) starve for fuel.

AIR CAPACITY

Because of its air-valve design, the Q-jet automatically adjusts air capacity up to its limit by engine demand. It can be tuned to run extremely well on engines ranging from 200—400 CID.

If you want maximum capacity for a performance engine, two sizes of carburetors have been available since 1971. The large ones have been used on various high-performance production vehicles, beginning in 1971 on Buicks. Pontiac started using these larger units in 1973. They will flow 800 cfm at 1.5-in.Hg vacuum. The majority of Q-jets used have 50 cfm less flow capacity.

Evaluate your performance requirements carefully before buying this larger unit to replace a regular Q-jet one. Don't be fooled by idle shop talk into believing you need extra carburetion. In many stock and some modified classes where the engine never reaches super-high rpm, it's likely to deter low- and mid-range torque.

SECONDARY SYSTEM

First of all, the air valve is designed to function only when capacity demands are fairly high (acceleration or high-speed running). Let's assume you are accelerating with a fairly heavy throttle. There is approximately 75° of primary throttle opening from idle to WOT. At approximately 35°—40° of primary opening, the secondary throttle blades begin to open. Regardless of your speed or rpm, a mechanical linkage continues to open the secondaries as the primary is pushed on to WOT. The geometry is such that during the

Secondary feed holes (arrows) above throttle blade. For quicker secondary-fuel response, hole can be located just below blade. Use care when plugging and redrilling.

last 35° of primary opening, the secondaries open from a closed position to their WOT position.

The air valve is independent of the primary or secondary throttle blades. It doesn't function until sufficient engine rpm, combined with secondary throttle opening, creates adequate depression under the valve. In short, the air valve doesn't open until the engine needs the capacity.

Many Q-jets have a built-in system that provides a shot of fuel as the air valve starts to open; a secondary pump shot, so to speak.

If your carburetor has this, you will see a hole near the leading edge of the air valve. Many of these feed holes are *above* the blade; some Q-jets have the feed hole just *below* the leading edge.

If your carburetor does not have this system, and it is to be used on a special performance application, other than street stock, you may want to add it. This requires getting a suitable used air horn for the required fuel pickup tubes.

Here's how it works: Extending down from the visible external holes are 1/8-in.-diameter fuel-pickup tubes. They extend into small reservoirs of fuel (one on each side of the fuel bowl). As the secondary throttle blades are opened, manifold vacuum is admitted up through the bores to the lower side of the air valve. This pulls an instant shot of fuel if the holes are *below* the blade. If the holes are *above* the blade, the shot comes as the air valve starts to open (as it exposes the holes to vacuum).

Once the fuel reservoirs are depleted, it takes a few seconds for them to refill. Like an accelerator-pump well, there is a contained volume and a refill requirement after use.

PREPARE FOR TUNING AND MODIFYING

Required Tools—Start with patience! You need more than the feel in the seat of your pants and the speedometer in the car to do serious tuning. Specific tools are required, but you'll also need a methodical approach to the project. If you have no intention of spending the time to be serious about tuning, run your carburetor as it comes out of the box and leave your tool box closed.

A vacuum gage, fuel-pressure gage, stopwatch and tachometer are essential. For regular competition, an air-density

Fuel-feed tube can be modified by drilling four holes through both walls to create smoother transition when secondaries are activated.

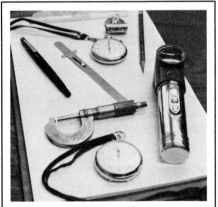

Tools for tuning include patience and hand tools, plus: stopwatch, one-inch micrometers, six-inch scale, inspection light like AC unit shown, clipboard, pencil and paper.

gage can be especially helpful. You'll also want a 1-in. open-end wrench, preferably the MAC S-141. It's designed for the fuel-inlet nuts on most carburetors.

The stopwatch can be particularly helpful. Pro stock racers often test acceleration over an initial 60 feet to refine starting techniques, tire combinations and carburetion. Most drag races are won in the critical starting period and by the initial acceleration over the first few feet.

Spare Parts—It is wise to have spare gaskets, an inlet-valve seat assembly in the size you use, an accelerator pump and miscellaneous screws and clips.

Jets and Primary Rods—Have these available for precise metering adjustments. If you're preparing a street rod, jets in sizes 068—073 will satisfy 80% or more of existing Q-jets and Dualjets.

If preparing for full competition, jet sizes from 072—076 will satisfy 80% of your Q-jet needs. Dualjets are out of the question for such efforts.

Primary metering rods in sizes 039—042 will do nicely for most street rods. Full competition requires rods in sizes 036—040. *Do not interchange* primary metering rods whose part numbers begin at 70 with those starting with 170.

Tuning Procedure Tips—It usually doesn't occur to some would-be tuners, or even some experienced ones, that changes must be made *one at a time*. It's too easy to be tempted, especially when you are sure that you need a heavier flywheel, different gear ratio, other tires, a different main jet, two degrees more spark advance and a different plug heat-range. Changing one of these at a time would be just *too slow*.

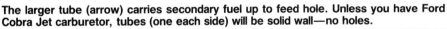

The larger tube (arrow) carries secondary fuel up to feed hole. Unless you have Ford Cobra Jet carburetor, tubes (one each side) will be solid wall—no holes.

Don't believe it. Remember, patience is a necessary ingredient. When you are trying to tune your carburetor, a complex piece of equipment, there's all the more reason to make one change at a time and then check each change against known performance data. Anytime you change more than one variable, you no longer have any idea which change helped, or hindered, the vehicle's performance.

Rejetting the carburetor to a specialized calibration that you heard or read about can become a tiresome task. Take this effort on if you're preparing for full performance and have the determination to follow through.

RPD spent plenty of time and money getting the calibration correct, and there is good reason to believe that they used more engineers, technicians, test vehicles, dynamometers, flow benches, emission instrumentation, and other equipment and expertise than you may have available for your tuning efforts. Remembering this may save your time, effort, money and temper.

When you have made jet changes from standard, use a grease pencil or a marking pen to mark the carb with the sizes of the main-jets you've installed. Some tuners use pressure-sensitive labels on the carb for noting which jets are installed. Others write the information on a light-colored portion of the firewall or fenderwell where it won't be wiped off during normal tuning activities such as plug changes.

Noting what jets have been used saves a lot of time when tuning because no time is wasted in disassembly and reassembly to see what jets are in the carburetor. Even the best memories are guaranteed to fail the jet-size memory test. Staggered jetting multiplies the problem.

Set mechanical fuel level where you want it before starting. For most performance use, 1/4 in. is the best mechanical setting.

Check throttle linkage and opening by removing the air cleaner (temporarily) and peering into the air horn. Have an assistant push the accelerator pedal to the floor as you check with a flashlight to make sure that the throttles open fully (not slightly angled). If they're not, figure out why and fix the problem.

Don't rely on checking the linkage by opening it by hand at the carburetor. That method doesn't take into account wear or misadjustment in the entire linkage.

Anytime you remove and replace the carburetor, check again to ensure the throttles are opening completely. It's easy to overlook and the cause of a lot of lost races or poor times.

Q-JET PERFORMANCE MODIFICATIONS

These modifications should be done on vehicles used only off-road, and they mainly apply to pre-'81 Q-jets. The final section in this chapter, "Performance Tuning

Electronic Q-jets," includes suggestions for increasing the performance of microprocessor-controlled Q-jets (post-'81).

The recommendations are based on my competition experience, but remember that your application may yield different results. My goal in this section is to supply you with information about modifications to your Q-jet that yield more performance per dollar spent.

ACCELERATOR PUMP

The ability of the engine to come off the line "clean" indicates a pump "shot" adequate for the application. A common complaint is, "It won't take the gas." Actually it's the other way around because the engine is not getting enough gas and a "bog" is occurring. Two symptoms often appear.

The first of these is that the car bogs, then goes. This can be caused by pump-discharge nozzles that are too small, so not enough fuel is supplied fast enough. The second symptom is one of the car starting off in seemingly good fashion, then bogging, then going once more. We are talking about race starting here, of course.

Solving the first problem may mean enlarging the pump-discharge nozzles, which may subsequently lead to increasing the pump stroke. In the second case, the pump-discharge nozzles may be the correct size, but the pump is not big enough to supply sufficient capacity.

As a general rule, the more load that the

Check float setting with scale or other appropriate gage before tuning.

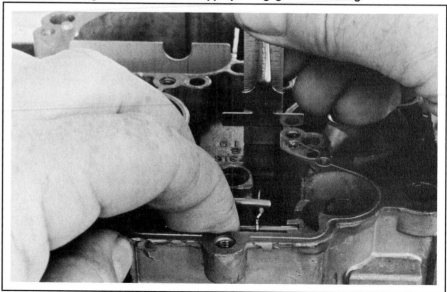

Herb Adams' Fire Am hustling around road course. Photo courtesy Herb Adams VSE.

engine sees, the more fuel that is needed in rate and volume. For instance, if an 1800-rpm stall-speed converter is replaced with a 3000-rpm converter in an automatic transmission, the *shooter (discharge nozzle)* size can be reduced. The same would be true when replacing a light flywheel with a heavier one. This is because the engine sees less load as the vehicle leaves the line in each instance.

Valve timing affects the pump-shot requirement. Long valve timing (duration) and wide overlap create a need for more pump shot than that required with a stock camshaft.

Carburetor size and position also affect pump-shot requirements. More pump shot is needed when the carburetor is mounted a long way from the intake ports, as on a plenum-ram manifold or a center-mounted carburetor.

The larger the carburetor flow capacity in relation to engine displacement and rpm, the more need for a sizeable pump shot to cover up the "hole" caused by slamming the throttles wide open. This is especially true with mechanically operated secondaries. The primary side of the Q-jet is small and efficient so the pump is generally quite adequate with minimum alterations.

Accelerator pump cutaway. Spring retainer (B) fits air-horn recess (A) to limit pump's upward travel. Some careful tuners move retainer down to gain added stroke, but avoid locating pump-cup lip above fill slot. (C) is duration spring. (D) is pump-return spring.

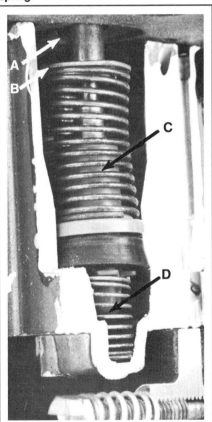

Plenum-ram manifold usually requires greater accelerator pump capacity.

When a carburetor is changed from one engine to another, the engine is changed into a different vehicle, or drastic changes are made in the vehicle itself, work may be needed to get the pump system to perform as you'd like. The main problem will usually be tip-in performance (opening the throttle from idle).

Shooter-Size Tuning—This is best done by increasing the nozzle diameter (or decreasing it) until a crisp response is obtained when the throttle is snapped open on an engine with no load. When crisp response is obtained, increase the nozzle size another 0.002 in. and the combination will usually be ready for racing.

Make any changes in small increments, because if you drill too large a hole, it will have to be plugged with epoxy and redrilled. Don't change the angle of the hole, it points the pump shot. Don't overlook the pump-duration springs. A slightly stiffer duration spring will also give you a quicker responding shot.

Pump-Shot Duration—The shot can be timed by shooter size, pump duration and return springs, pump capacity and pump rod to lever setting. The duration spring on the shaft is a safety valve that compresses when the throttle is slammed open. The compressed spring force against the pump cup establishes delivery pressure in the pump system. Thus, delivery rate depends on both system pressure and shooter size.

The duration spring must never be adjusted, or of a wire size, that keeps it from being compressed. And, don't replace the spring with a solid bushing in hopes of improving pump action. Either course of action will net you a badly bent pump linkage because gasoline won't compress. Something has to give or break if the throttle is slammed open and there's no way for the system to absorb the shock.

Pump Capacity—On Q-jets, pump capacity can sometimes be modified by lever/linkage changes to get maximum pump stroke. The pump piston should bottom in its well. Stroke increases may be gained by lifting the piston to a higher starting position, but not past the fill slot or pump-well entry. Each carburetor must be examined to see whether *any* changes can be made.

The limit of pump travel in a Q-jet pump bore is 21/32 in. from the bottom of the cup to the lower edge of the fill slot. Anytime the pump cup rises above the lower edge of

Underside of air horns reveals one with counterbore (arrow) for pump-rod retainer. Other, which is flat, may not allow same travel. Note this difference before modifying to get maximum pump travel.

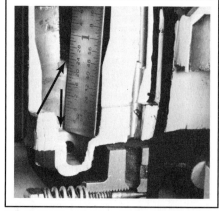
Fill-slot bottom (arrows) can sometimes be raised about 1/16 in. for more pump stroke. This is easiest when pump is bored for large 0.730-in. cup

Accelerator pump well being bored to 0.730-in. diameter for larger pump cup. Polish bore with crocus cloth to ensure seal between pump cup and bore.

the fill slot, the pump action will be inefficient. This is because a portion of the pump travel will be wasted while it expels fuel through the fill slot into the bowl, resulting in an intolerable pump lag.

It has been suggested by tuners that 3/32 in. can be ground off of every Q-jet pump rod to improve its function. In my experience, some early models can be adjusted for the full 21/32-in. travel by grinding (or filing) a mere 1/32-in. from the rod top. I have not found any units that allowed more than 3/32 in. to be ground off the rod end without raising the cup above the fill slot.

Capacity can also be increased by raising the bottom of the fill slot by up to 1/16 in. Care must be exercised so the cup lip does not end up above the slot when the pump assembly is up. With this modification, the duration spring can be shimmed to a shorter length (compressed). This allows the assembly to lift higher before it is stopped by the air-horn casting. And, it changes the spring characteristic so that the pump shot is discharged faster.

Another method for increasing capacity is to enlarge the pump-well cavity and use a larger pump cup. The pump well must be bored to a larger diameter by a competent machinist. The stock Q-jet pump well has a 0.640-in diameter and the stock pump rubber diameter is about 0.645 in.

You can increase the capacity by more than 30% by using the 0.735-in. diameter pump cup from a late-model big bore (1-1/2 in.) Rochester two-barrel pump assembly. Bore and polish the Q-jet pump well to 0.730-in diameter, then install the two-barrel pump cup. These are available from Delco or as GM parts numbers:

7036282 Chevrolet, Pontiac, Buick
7046028 Oldsmobile
7037562 Oldsmobile
7037561 Pontiac

Pump Travel—This is easily checked on an assembled carburetor. Use a small screwdriver to push the pump to the bottom of its bore. When it is at the bottom, scribe a mark on the pump rod even with the air horn. The small amount of the rod protruding above the air horn may be as little as 1/32 in. in some cases.

If so, the maximum to be removed from the end of the rod cannot exceed 1/32 in. because the pump linkage cannot push the pump rod below the level of the air-horn casting. In fact, grinding off more than this could *reduce* the pump travel if the rod top

Stock Q-jet pump cup at left. Cup at right (0.735-in. dia.) is from large 2G.

When checking travel, coat tip of pump rod with machinist's dye. Then, scribe mark to indicate lowest point of travel.

With pump at bottom of its bore, measure exposed portion of pump rod.

Next, measure exposed portion of pump rod at top of travel. Difference between this and bottomed measurement is total pump travel.

Most influence on heavy or WOT fuel requirements is made by changing secondary metering rods. Simply remove air cleaner and one rod-hanger screw. Power piston modifications are more difficult.

Adjust pump rod by bending main rod, not end. Arrow indicates broken end caused by bending. End is hardened and straightening breaks it.

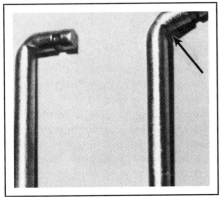

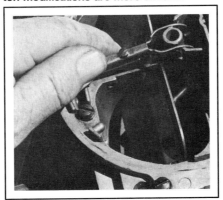

ends up below the top edge of the bore in the air-horn casting.

Pump Linkage—Preload the linkage at idle so there is no slack. The pump lever must push down on the rod slightly. Any movement of the throttle must continue the pump's downward movement in its bore. I prefer to make this setting so the linkage pushes the pump rod about 1/64 in. down in its bore. This ensures the pump-cup lip is just slight *below* the fill slot. The most you will have to do here, once you have the 21/32-in. stroke arranged, is to bend the actuating rod to get this preload.

If the accelerator-pump linkage on the carburetor requires a pin to be driven out of the lever to get the air horn off, consider installing the pump link from an old Q-jet. Replace the actuating rod with the one requiring a C clip. This will make linkage changes on the carburetor simpler and faster. Q-jets produced since 1981 have heavier pump rods that can't be replaced with the older ones.

POWER SYSTEM

You may immediately think, "I know just what to do—take it out!" Regardless of all information to the contrary, *there is seldom any need to deactivate the power system.*

Power-system tuning requires the use of a vacuum gage. Let's take the example of a car equipped with a camshaft that provides such a low manifold vacuum at idle or part-throttle, the power system operates, or perhaps turns on and off due to vacuum fluctuations.

In this instance, a power system that opens at a still lower vacuum is needed. If a 7-in.Hg manifold vacuum power system is in this carburetor, the vacuum could drop to 5 in.Hg at idle or part-throttle. The power system actuating spring will need to be weakened or shortened so it actuates the system at 4—5-in.Hg vacuum.

You can see that it is essential to know what the manifold vacuum is at idle. It is also one measurement case where it is necessary to have a gage that is not highly damped. That is, the needle will have to "jump" to follow the vacuum fluctuations or you won't know how low the vacuum is getting.

Another application that demands using a vacuum gage is racing in a class where carburetion is limited to a certain carburetor type or size that is really too small for the engine. A class demanding the use of a single two-barrel would be typical. In this case, the vacuum in the manifold may remain fairly high during the race.

The power system should always have a higher operating point than the highest manifold vacuum attained during racing. If this is not the case, the power system will close and the engine will run lean. Disaster may result, usually in the form of a holed piston. Sometimes the power system will get lean enough to cause "popping" sounds from the exhaust.

For example, a carburetor altered to have a 3.0-in.Hg vacuum power system will go lean if the manifold vacuum is 4.0 in.Hg through the high-speed portion of a course.

Change The Operating Point—Modifying the operating (cut-in) point of the power system requires changing the power-piston spring on the Q-jet, or the power-piston assembly on the 2G. The accompanying tables show springs and assemblies for making changes. When different power-piston springs or assemblies are not easy to get, the change can be accomplished by removing two to three full coils of the power-piston spring.

Power Piston Assemblies for 2G 1-1/2-inch carbs

Assembly Number	Cut-in point (in.Hg)
7035603	11-8-6
7006323	9-7-5
7009718	8-6-4

Power Piston Springs for Quadrajet carbs

Assembly Number	Cut-in point (in.Hg)
7037734	14-4
7032758	11-5
7037305	10-6
7036019	8-4
7029922	7-3
7037851	6-3

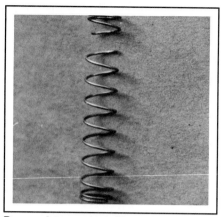

Power-piston spring with 1-1/2 coils cut off. Count 1st or end coil as 1.

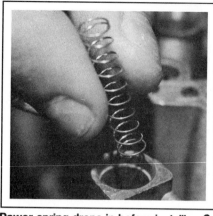

Power spring drops in before installing Q-jet power piston. Make sure cavity bore is smooth and free from scratches. Polish vacuum piston to ensure it operates without sticking.

Observe and check the cut-in point of conventional power systems with an adjustable regulated vacuum source. This is usually a feature of a distributor-testing machine. Or, vacuum can be tapped off the manifold of an idling engine.

In either case, a vacuum gage must be included in the plumbing so you can see when the power system starts to operate and when it is fully actuated. The vacuum source must be applied to the base of the carburetor, which may require making an adapter plate to apply the vacuum without leaks. You can't use a hand-operated vacuum pump for this measurement because its volume isn't adequate.

For comparison purposes, let's say you want to modify a 2G or 4GC. Remember that the power piston is held *up* by manifold vacuum and the spring tries to overcome this vacuum. As vacuum is decreased, the spring overcomes the vacuum force so the actuating rod is driven down against the power-valve pin to turn on power-system fuel.

Start with about 10—12-in.Hg vacuum and reduce this so the power piston actuating rod descends to move the center pin of the power valve ever so slightly. Record the vacuum that causes this to occur, as this is where the power system starts to operate.

Then decrease the vacuum still further until the rod end bottoms against the power valve, indicating that the valve is fully open. This figure indicates the vacuum at which the power system is fully on.

On the Q-jet, the power piston is held *down* by manifold vacuum, and the spring tries to overcome this vacuum. As manifold vacuum decreases, the spring overcomes the vacuum force, raises the piston and pulls the metering rods up so their minimum diameter (power tip) is in the primary main-metering jets. Power-system fuel is now available.

The Q-jet power system should always be fully operational by 3—4-in.Hg vacuum. If not, then the spring is too weak, or you have cut too many coils off the power-piston spring.

Although power-piston and metering-rod operation is most easily observed with a cutaway carburetor, a feeler can be inserted through the forward vent of a Q-jet so that it lightly rests against the top of the power piston.

The feeler (soda straw or match stick) should be stiff enough so you can use it to push the power piston down, representing the situation when vacuum is applied, so the largest portion of the primary metering rods are inserted into the primary jets. Mark this position on the feeler. Also mark the full-up position—the spring is holding the piston in full-up or power-system-on position.

Apply vacuum at about 10—12-in.Hg and reduce this so the power piston raises. Record the vacuum at which the piston starts to raise and the vacuum at which the piston reaches its full-up position. The first value will approximate the start of power-

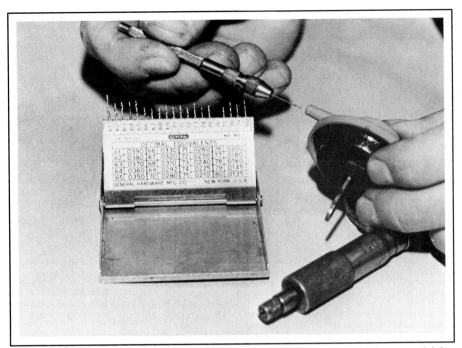

Wire drill index with drills from 0.0135—0.039 in. (numbers 80—61) is essential for serious tuner. Pin vise holds drill for turning by hand.

system operation. But is difficult to tell whether this is really the start unless you can observe the position of the tapered portion of the metering rods in the main jets. You can rig a cutaway of an old carburetor to observe the rods' start.

TUNING FOR RACING CAMSHAFTS

Radical camshafts sometimes require carburetion changes to get satisfactory operation below 2000 rpm. A racing camshaft with lots of valve-timing overlap can cause seemingly insurmountable carburetion tuning problems. It takes some extra effort to ensure that the engine will idle at a reasonable speed and will not load up the plugs when driving warmup laps, caution laps or running back to the pits after a drag run.

Preliminary Inspections—Check the accelerator-pump setting and make sure the bowl screws are tight. Look at the underside of the carburetor with the throttle lever held against the curb-idle stop (not against a fast-idle cam). Note the position of the primary throttle plates in relation to the transfer slots or holes. This relationship has been established by the factory engineers to give the best off-idle performance.

Baseline Measurements—Check the clearance between the throttle plate and bore with a feeler gage or pieces of paper as you hold the throttle lever against the curb-idle stop. Note this clearance in your tuning notebook. Record every modification as you proceed, regardless of how good your memory may be.

Install the carburetor and then start the engine. If you have to increase the idle-speed setting to keep the engine running, note how many turns or fractions of a turn, are needed to open the throttle to this point. Adjust the idle-mixture screws for the best idle. If the mixture screws do not seem to have any effect on the idle quality, note that fact.

Use a responsive (not highly damped) vacuum gage to measure the manifold vacuum at idle. If the engine idles with the manifold vacuum occasionally dropping to a value lower than that required to open the power valve, you will need to install a valve that will remain closed at idle before proceeding.

Precautions—Because you will be drilling very small holes, requiring a "wire drill" set, you must proceed in very small increments. Even a 0.002-in. increase in an idle-feed restriction of 0.028 in. increases area and flow by 15%. It is easy to drill holes, but there is no easy way to get back to the starting point. Zealous drilling may require a new throttle body.

Wire drills, incidentally, are not used in a power drill but must be held in a pin vise. Pin vises are available where you buy wire drills—at precision tool supply houses or model shops.

An adequate accelerator-pump setup usually eliminates any need to work on the off-idle transfer fuel mixture, which is controlled by the idle-feed restriction. The mixture can be checked by opening the throttle with the idle screw until the main system just begins to start, then backing off the screw until it just stops. If idle and off-idle performance turn out to be unacceptable, increase the idle-feed restriction slightly.

Modifying Throttle Plates—Take the carburetor off the engine. Turn it over and note where the throttle plates are in relation to the off-idle (transfer) slot. If you can measure more than 0.040 in. of the slot between the throttle plate and the base of the carburetor, drill a hole in each of the primary throttle plates on the same side as the transfer slot. If holes already exist in the throttle plates, enlarge these holes.

Small-displacement engines require smaller holes than big-displacement ones. Start with a 1/16-in. drill on your first attempt and then work up in 1/32-in. steps. Before reinstalling the carburetor, reset the idle to provide the same throttle-plate-to-bore clearance that you made as a baseline measurement.

Start the engine. If the engine idles at the desired speed, the holes are the correct size. Too slow an idle indicates that the holes need to be larger; too fast an idle indicates the need for smaller holes. If holes have to be plugged and redrilled, either solder the holes closed or close them with Devcon "F" Aluminum.

When you have the holes at the correct size, which may require the use of number or letter drills to get the idle where you want it, note the size. Where there are two throttle plates on the primary side, both plates should have the same size hole.

Modifying Curb-Idle Discharge Ports— Do the idle mixtures screws provide some control of the idle quality? That is, do they cause the engine to run rough as the idle needles are opened? If not, check the size of the curb-idle discharge ports into the throttle bore.

You can check this by unscrewing a mixture needle and using your number drill set. If the port is not within 0.030 in. of the largest part of the needle, say a 0.095-in. port with a 0.125-in. needle, drill out the discharge ports for each barrel equipped with a mixture screw.

NOTE: Carburetors on emissions-controlled engines may have small curb-idle ports. *Before* increasing the idle-feed orifice to make the carburetor work with a racing cam, always open the curb-idle ports to their size before emissions control. This is about 0.095 in. for the Q-jet.

Modifying Idle-Feed Restrictions—If this does not give you idle control, then open the idle-feed restrictions approximately 0.002 in. at a time until some control is achieved. Correct control is indicated when the engine runs as smoothly as possible, and turning the idle-mixture screw either way causes rpm to drop and roughens the idle as the mixture is leaned or richened.

PRIMARY THROTTLE CLEARANCE
Never back off the idle-setting screw or change the fixed or spring stop to allow the *primary* throttle(s) to seat against the throttle bores. Always maintain the original clearance between the plates and bores. Secondaries will be seated against the bores. This is correct and normal practice in RPD carburetors.

DRAG RACING MODIFICATIONS

With the growing popularity of the American Hot Rod Association (AHRA) showroom stock and National Hot Rod Association (NHRA) stock classes, which require retaining the original-equipment carburetor, it has become necessary to make do with what you've got.

In the case of almost all late-model, four-barrel-equipped GM V8s and V6s, this means using the Q-jet. The popularity of RPD carburetors in many racing applications trails other makes because the factory doesn't pursue that market.

Nevertheless, the performance modifications specified in the previous sections of this chapter transfer well to drag racing applications. Included here are specific tips for applying the Q-jet to this type of high-performance driving.

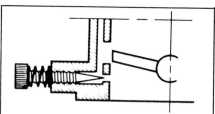

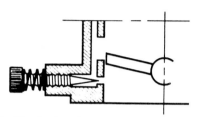

A—Transfer slot in correct relation to the throttle plate. Only a small portion of the slot opens below the plate.

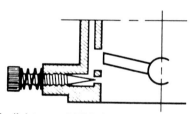

B—Throttle plate closing off slot gives smooth idle with an off-idle flat spot. Manifold vacuum starts transfer-fuel flow too late. Cure requires chamfering throttle-plate underside to expose slot as in **A**.

C—If slot opens 0.040 inch or more below throttle, rough idle occurs due to excessive richness and little transfer flow occurs when throttle is opened. A long flat spot results.

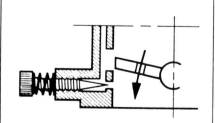

D—Correcting condition in **C** requires resetting throttle to correct position **A** and adding hole in throttle as described in text.

Inlet System—Delco has three sizes of needle valves available for the Q-jet: 0.110-, 0.125- and 0.135-in. diameter. While I don't automatically recommend discarding the stock needle valve in favor of the largest available, some applications might benefit by swapping the existing fuel inlet size for a larger one offered by Delco or another supplier.

The needle is opened and closed by a closed-cellular plastic float that should be examined carefully for scratches or other irregularities that could affect its operation. Having a spare one on hand is wise because they are subject to damage, and at a race it's tough to get replacements.

Primary Fuel Metering—For WOT operation, meter the primary and secondary WOT fuel requirements *first*. If this causes part-throttle richness, increase the part-throttle metering-rod area to reduce part-throttle fuel flow. Use secondary metering rods with short power-tip ends so the intermediate opening power positions are lean.

Significant metering changes for drag racing will have to be accomplished by changing the jets instead of the rods. Part-throttle mixtures can become excessively rich and premature plug fouling could become a problem if the vehicle is subjected to slow driving in the pit and staging areas.

In such cases, a larger metering rod would be required to achieve correct part-throttle metering and yet retain the richer WOT fuel. Cars running with automatics

Stamped numbers and letters on primary metering rods are small. Spark-plug inspection light is useful for checking all small parts.

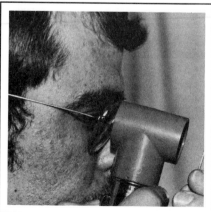

Super Stock H Automatic wrinkles rear tires running with original equipment carburetor: Q-jet.

using low-stall-speed torque converters could also benefit by metering rod changes. They would be sluggish coming out of the hole if the low-end metering is rich.

Secondary Fuel Metering—This control is provided by replaceable rods because the fixed-size, 0.135-in.-diameter orifices are not removable from the bowl.

The secondary metering rods are attached to what is referred to as a *hanger*. Rochester makes 20 types of hangers. They differ in the location of the hole where the secondary rods attach. These hangers are identified by a single stamped letter. Hangers B through V have a hole location change of 0.005 in. per letter.

Change the hanger hole location to alter the relative depth position between the rod and the secondary metering orifice. Consequently, this changes the restriction to fuel flow. Delco offers only one replacement hanger: Part No. 7034522, identified with the letter V. RPD uses the 20 different hangers to do specific model calibrating.

So, you have another reason to buy inexpensive, used Q-jets. That's the only way to build a stock of different hangers. Roe, Inc. can supply three different hangers for those who want an additional edge: lean (0.520 in.), average (0.570 in.) and rich (0.615 in.).

There are a few modifications explained below, which can be made to the secondary fuel delivery system for quicker and more consistent engine response at WOT.

h DIMENSION	IDENTIFICATION STAMP	h DIMENSION	IDENTIFICATION STAMP
.520	B	.570	L
.525	C	.575	M
.530	D	.580	N
.535	E	.585	O
.540	F	.590	P
.545	G	.595	R
.550	H	.600	S
.555	I	.605	T
.560	J	.610	U
.565	K	.615	V*

*Only service hanger, P/N 7034522

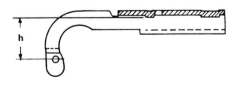

Q-JET SECONDARY METERING ROD HANGERS

If stock hanger has secondary rod hole on high or low side, carefully drill another hole and have rich and lean hanger in one.

Secondary rods are inserted with bent ends pointing into hanger (arrows).

Top view showing upper vertical passages to drill. Tool points to one, arrow indicates other.

Cutaway detailing fuel passage (A) with 0.160-in. diameter intersecting with vertical passage for secondary air-bleed tubes. Small lower passage causes restriction (B) at intersection. Drill out (A) to 0.197 in. Left passage has been drilled.

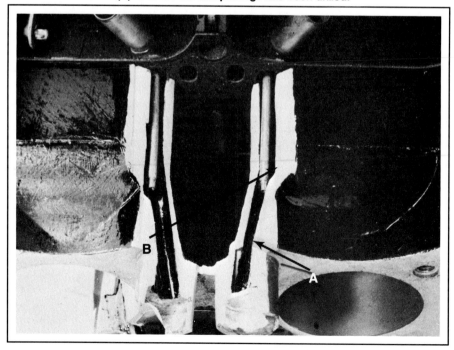

Enlarge fuel passages in the secondary system to increase its capacity to deliver more fuel. Secondary fuel passes through the metering orifices to a small chamber below the float bowl. At this point, the fuel travels upward through two 0.160-in.-diameter passages that intersect with the 0.197-in.-diameter passages leading to the nozzles. The two smaller-diameter passages sometimes are an unwanted restriction to high fuel flow.

This restriction can be changed by drilling the passages out to 0.197 in. to match the adjoining passages they feed. Refer to the accompanying photos to see how this is done.

With the main body of the carburetor inverted, you will note two plugs side-by-side in the center of the carburetor. To get at the 0.160-in. passages referred to above, these plugs must be removed. On late-model Q-jets these plugs are die cast and spun in. If you have replacement plugs, drill out the originals.

Some early Q-jet models have brass cup plugs that can be removed by prying them out with a good-quality ice pick or awl. Care should be taken during this procedure not to do any unnecessary damage. Replacement plugs are available at many auto-parts stores.

Drill upper passages with bit *not larger* than 0.200 in. to depth *not exceeding* 1-1/4-in. If bit breaks through to venturi (arrow), it must be repaired.

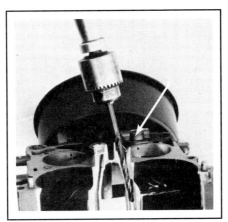

Remove fuel-chamber plugs (arrow) before enlarging lower channels. Using bit *not larger* than 0.200 in., drill lower channels to intersection with upper passages.

Late-model Q-jet fuel-chamber plugs are driven in and diecast material is peened firmly over plug as shown. These plugs don't leak, put they aren't extracted easily either.

Remove retaining metal from top of plugs. Start with small chisel and care.

Next, grind off plug's lip with rotary file. Keep firm grip on tool and don't remove any extra metal.

More heavy machinery: cutting stone being applied to plug lip. Keep firm grip on tool and remove only lip.

Note thickness of removed plug held in hand. Second plug being drilled so it can be grasped and removed.

Just getting warmed up now. RPD didn't intend plugs to fall out or be pulled out.

Early cup-type secondary-well plugs usually can be removed by catching inside wall with sharp awl or punch and prying out. This pair secured by epoxy.

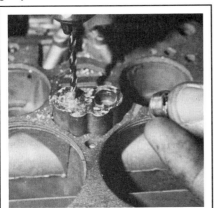

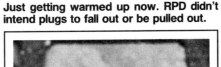

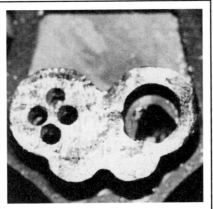

These plugs have caused metering problems by leaking, so I recommend using a good epoxy to cement them into the well cavities. Don't treat this problem lightly. This is the extreme botton of the fuel reservoir and a leak here will let approximately a teacup of fuel into the engine. Early Q-jet models suffered most from this leakage; models since the '70s seldom leak.

Move internal air bleeds to change fuel restriction in the secondaries. By inverting the top of the carburetor, you will see four tubes, each approximately 1-1/4-in. long. On a few Q-jets there will only be two. The two smaller tubes, approximately 1/16-in. in diameter each, are air bleeds to the secondary fuel-delivery passages. The other two tubes, about 1/8-in. diameter, feed fuel.

When the carburetor is assembled, the tubes extend into the intersection where the previously discussed passages meet. By tapping these tubes farther into the air horn—approximately 3/16 in.—the passage restriction can be reduced. Their standing height will be 1 in. when driven in to the recommended depth. They do not move easily, so fabricate a piece from aluminum or bronze that will transmit driving forces to the tube's shoulder without damaging the orifice at the tip.

Drill internal fuel tubes to create a smoother transition when the secondaries are activated. The 1/8-in.-diameter tubes act as a secondary accelerator pump by delivering a shot of fuel before air velocity is high enough to draw fuel from the secondary discharge nozzles.

The tubes extend into wells at the rear of the float chamber (right and left side). The wells receive their supply of fuel through a 0.028-in. diameter orifice located about 3/4 in. up from the floor of the carburetor's float chamber.

When the secondary throttle blades start to open, a low-pressure area is created beneath the secondary air valve. Fuel flows from the wells, up the tubes, and discharges from the 0.060-in. drilled orifice located just beneath the secondary air valve. On some carburetors, this discharge orifice will be located just above the air valve, which means the fuel shot will be delayed until the air valve just starts to crack open.

To create a smoother transition when the secondaries are opened, the 1/8-in. tubes should be drilled. Starting approximately

Make tool to drive air bleeds by drilling aluminum tubing to 29/32 in. with 3/32-in. bit. Then counterdrill another 1/4 in. with 1/16-in. bit for clearing metering tip.

1/4 in. from the end of the tube, drill four 0.030-in.-diameter (1/32-in.) holes about 3/32-inch apart straight through both sides of the tubes.

Increase fuel flow into the float-chamber wells to speed their refilling. They require about 15 seconds to refill through the 0.028-in.-diameter orifice just off the float-chamber floor. If after one of two hard burn-outs, you encounter a hesitation or bog, then the wells may not be replenishing adequately. Increasing the size of the drilled orifice could solve this.

Keep in mind the refill orifice serves as a metering restriction during extended secondary use. Secondary fuel metering could be affected drastically if this orifice is punched out too large. Make modest 0.001—0.002-in. changes.

Alter the secondary air valve opening rate for more performance. It is opened by the low pressure created when the mechanically operated secondary throttle valves are opened. But the air-valve opening rate is not instantaneous. If your engine is race-prepared, then it will benefit by getting the air valve to open quicker.

Tubes (arrows) have been driven to preferred standing height of 1 in.

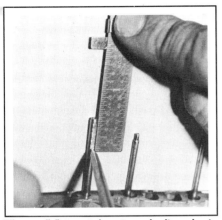

File small flat on tubes to make it easier to mark and prick-punch for drilling.

Hold air horn as illustrated to work on tubes. Quality awl is excellent tool for light prick-punching.

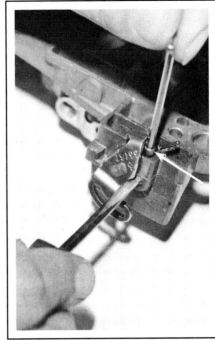

Adjust secondary air valve on early Q-jet by inserting screwdriver in slot on shaft side and loosening allen screw (arrow). Turn screwdriver counterclockwise until air valve is closed with zero tension.

Special chuck and shaft are used to hold 0.030-in. bit. Drill press is better for this fine work, but steady hands and care will work.

Hand-held drill job may net some visible irregularities, but if hole sizes and spacing (arrows) are correct, it's fine.

Note precise factory-drilled holes. These tubes were standard in 1970 Ford Cobra Jet Q-jets. Sorry, tubes aren't replaceable item so can't be purchased separately.

Auxiliary feed tubes get fuel from reservoirs (A). (B) indiates 0.028-in. hole that fills left side. Similar hole is in right side.

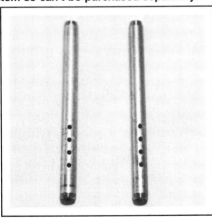

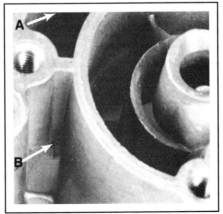

Two major control mechanisms regulate its opening rate. First, on the right side of the air-valve shaft is an adjustable spring applying tension against the air valve. The factory settings are usually between 1/2 and 3/4-turn tension. Reduce this tension in steps of about 1/8-turn at a time to change the opening rate.

The quicker the air valve opens, the better for maximum performance. The limiting factor will be a lean sag at opening if it's too quick. Adjust to fit your application.

The second control mechanism is on the right front side of most Q-jets. It is an enclosed plastic vacuum diaphragm attached to the air valve by a linkage arm. This diaphragm has a dual purpose. First, it serves as a vacuum break for the choke. Second, it slows the air-valve opening.

Don't throw the diaphragm away because this will cause the air valve to open too quickly. When this happens, it will slam shut, bounce open, and lose you time.

If you remove the neoprene hose from the diaphragm, you will see a hole of approximately 0.010-in. diameter. By increasing this diameter in steps of about

Allen setscrew may be recessed as shown here (arrow). Air valve is open.

Don't turn screw more than one turn from zero position or secondary air-valve tension spring (shown here) will be distorted.

Modify stop (arrow) by grinding to allow full 90° opening of air valve. Some Q-jet valves open only 50°, which limits 750 cfm unit to 590 cfm.

Steel diaphragm (arrow) is difficult to alter because damping restriction is internal. Replace with plastic type before tuning air-valve opening rate.

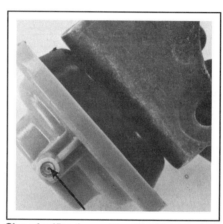

Size of orifice (arrow) is critical to premium operation of Q-jet. Make incremental changes and review performance after each.

Increase slot depth by cutting with hacksaw.

Next, release tension on *outer* spring by removing straight end of spring from its tang (arrow) with pliers. Hold secondaries closed while rotating lever clockwise to let spring clear its slot in shaft end.

First step in removing lever from secondary shaft is to pry edge of *inner* spring (left photo, arrow) off its ledge, then catch it with finger (right photo, arrow) and allow it to unwind.

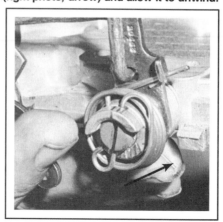

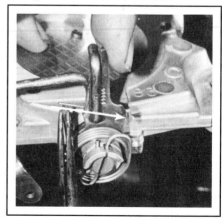

138

How far should you saw slot? First effort of 0.050-in. deeper moved 45° secondary pickup point to 38°. Further cutting (0.020 in.) and cleaning moved pickup point to 30°. I don't recommend secondary pickup quicker than 30° or they will be open all the time.

Mark lever by scribing file mark (left photo) to show original slot location before making any alterations. Line marking end of original slot dimension is shown in right photo (arrow).

Underside of throttle body shows over-center secondary blade position as result of modification.

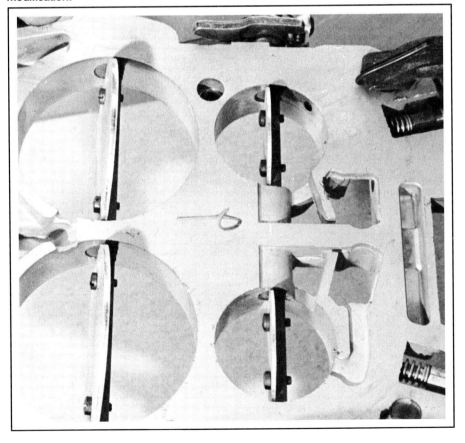

0.002 in., the damping effect on the air valve can be progressively reduced. All damping effect will be lost if the orifice is drilled much beyond 0.025 in.

Modify the secondary throttle linkage to get secondary throttles open faster. Another alteration can be done to quicken secondary opening: Open the secondary throttles sooner by modifying the linkage from the primary throttles. Of course this modification sacrifices economy because the secondaries will be working at almost any road speed.

To open the secondary throttles earlier requires modifying the secondary pickup lever. There are two types. In one case, a slot must be deepened so a spring wound around the secondary throttle shaft causes the linkage to act quicker.

In the other arrangement, the hole for the rod to the primary throttles must be relocated downward toward the secondary throttle shaft. Details for handling the first type mentioned are provided in the next section. It also explains how to obtain 90° or more secondary throttle position.

Secondary Throttle-Blade Angle— Some Q-jets open the secondary throttles to a full 90°. Others open the secondary throttles to a position just less than 90°. In either case, the carburetors will flow their full rated capacity. In some cases, notably Chevrolet V8s, the front cylinders have a tendency to run lean. This may be caused by the secondary throttles not being fully opened by the stock throttle linkage.

  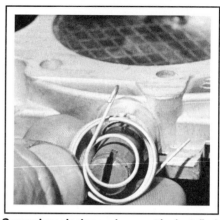

Check progress with protractor on primary throttle shaft (left photo). Set it up with wire pointer clamped to throttle body so 0° indicates closed-throttle position. Protractor shows 30° primary throttle opening from closed (right photo) and primary lever linkage is starting to open secondaries.

Secondary closing spring must be installed first when replacing lever. Note wrap on outer, hooked end. Hook must be attached to lever, then inner straight end tensioned.

 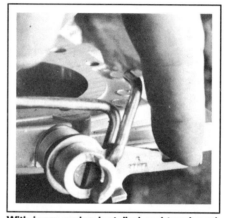 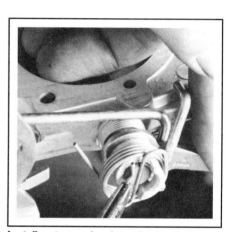

Spring has been hooked onto lever (arrow). Guide straight end of spring past obstacles with pliers. Put slightly more than one turn tension when spring end is placed on ledge behind secondary shaft.

Final twist on spring brings it behind tang as shown and lever is completely installed.

With inner spring installed and tensioned, hold throttles closed with one finger and rotate lever clockwise to position shaft slot and lever opening as illustrated.

Outer spring curved end is installed in lever slot and shaft end slot, and straight end of spring is being wrapped toward its tang.

Install outer spring into slot in shaft end, pushing lever toward throttle body as you do so.

Air gap between actuating rod and secondary lever should be about 0.005 in., as measured here with feeler gage.

  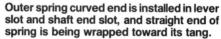  

This leaness can be reduced or cured in many instances by reworking the linkage so the secondary throttles open to a position past 90°, up to 5° farther, so the throttle blades deflect the secondary mixture toward the front of the manifold.

The accompanying photos and captions describe this modification. It is important to keep the vertical portion of the rod striking the opening tang on the secondary lever. It provides additional leverage for opening the secondaries against the friction tending to hold them closed in their bores.

The secondaries must close easily and decisively when the work is completed. This is a function of the closing tang on the primary throttle lever and of spring tension on the secondary closing spring.

Careful thinking and a studied approach are required to operate on either of the secondary-opening mechanisms—there are only two types. Proceed slowly and methodically, or be prepared to buy a new carburetor throttle body. The levers are not sold separately.

Adjust vertical part of actuating rod with large pliers to set air gap (arrow).

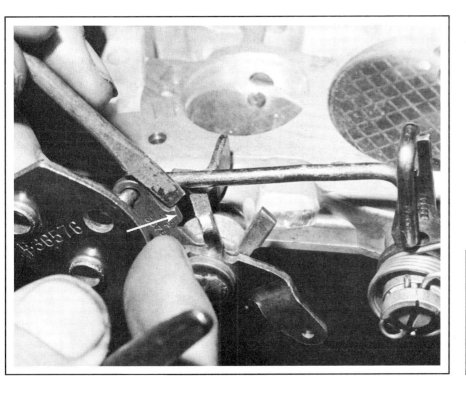

Positive secondary closing is lost after modifying pickup point. Changes move tang on outer primary lever away from inner secondary-actuating lever closing tang (note screwdriver). Weld material onto closing lever at arrow to close gap to zero at curb idle, if secondary closing is necessary. Submerge throttle body of Q-jets with teflon-coated shafts in water to keep them cool and work quickly if any welding is done on levers.

SECONDARY ACTUATING ROD
Don't bend the primary-to-secondary actuating rod except at the vertical end to adjust for 0.005-in. clearance with the tang atop the secondary lever. It may be tempting to bend the rod to *shorten* it, but don't! If you do so, the secondaries may not close, or may hang up over-center. You don't want that to happen.

Off-road racing subjects carburetor to extraordinary stress. It must be throughly prepared before rampaging in the rough. Photo by Tom Monroe.

OFF-ROAD RACING

Prepare the Q-jet well for off-road racing. It is subjected to jolts, bumps and extra stress. Controlling fuel level through these forces and motions is no light task. These modifications increase its durability and performance.

Floats—Floats have many shapes and sizes. Their size is largely influenced by the bowl, and so is shape to some degree, but generally there is latitude for subtle alteration to their design.

Each time you stab the throttle when racing off-road, fuel will stack up rearward and can hold the float so the inlet valve is closed. Leaning of the mixture results. Accelerating or climbing a steep grade adds to the problem.

If this driving causes lean-out, I would try to eliminate buoyancy at the outer end of the float. I have gone float pontoon hunting, and once the ultimate shape was found, it was integrated with the proper float arm. This practice is quite satisfactory when metal floats are used, because they can be soldered in place.

The material used in most modern floats is a closed-cell plastic, which is easy to alter, but not likely to last too long. The material is easily cut and shaped, and because of its cellular composition, it will not leak.

This is true to a point, but once the outside surface is cut, some of the exposed outside cells may connect to inner cells. In this case, fuel gets in and the float gets heavier and is less bouyant.

FLOAT LEVERAGE DETAILS

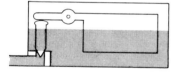

Regulated fuel lever, static condition.

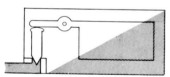

Acceleration with front-pivoted float. Turning with side-pivoted float.

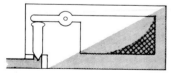

When fuel stacks away from the inlet needle, reduce the portion of the float that is most effective. On a long acceleration or corner, stacked fuel can hold the needle closed and you will get lean-out.

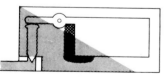

Float leverage is lost when the fuel stacks toward the inlet needle, so add more material to increase buoyancy. A float shape with more material in the fuel may help to shut off the needle (control fuel level). Otherwise, the mixture is apt to be rich. During an acceleration you might be able to get away with it. On a corner it is a no-no.

Top and side views of Q-jet floats show variety of shapes used to fit particular off-road applications.

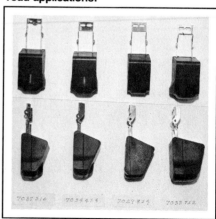

If you are running in rough terrain and it's evident you need a new shape float: cut it. If you anticipate such surgery, have a small can of clear shellac with you. After the shaping is done, put a thin, light coat of shellac over the new surface and let it dry thoroughly. Neither factory nor service dealers will recommend such practice because it can shorten the life of the float. It is strictly a special-application, do-it-yourself exercise. Carry an extra float or two just for insurance.

Remember if you cut away much of the lower portion of the float, particularly at the outer end, appreciable buoyancy will be lost. Readjust with additional mechanical measurement from the air-horn gasket surface to the float, to get back the liquid level you had before alterations. If you make small systematic changes when trying to beat one problem, you are less likely to create another one elsewhere.

The float arm leverage will have one of two dimensions. The one with a short-fulcrum float arm is best for off-road racing. I have made both types work under the worst of conditions, but the short fulcrum leverage makes tuning easier.

This sounds easier than it actually is because you can't simply install the short-arm float into a carburetor that originally had one with the long arm. You actually have to switch the entire carburetor. Short- and long-arm floats cannot be interchanged.

The 1965—66 and some 1967 Q-jets with the diaphragm-type fuel inlet assembly had the long fulcrum. Because they date back a few years, parts for modifying them will be limited. I recommend you avoid them for off-roading.

Fuel Filter—Use a good in-line filter just ahead of the carburetor. Plan to change it often, particularly if the carburetor sits days or weeks at a time between runs. Fuel systems can rust and gather condensation and harmful foreign contaminants when sitting. Like every other component, the filter has limitations you can exceed. If driving off-road, be on the safe side and change it regularly.

Inlet Valve—Review the section earlier in this chapter about modifying the inlet valve. To determine the best size for your off-road Q-jet, carefully check performance with a fairly large valve. In most cases, use the stock one.

Next, install an inlet seat with a smaller orifice. Repeat the exact performance

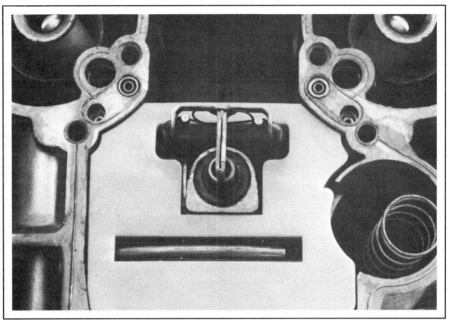

Plastic spacer limits fuel volume that can rush away from float during forward brake stop. Float is centered between primary bowls.

Short-fulcrum float (A) compared with long-fulcrum float (B) shows difference in leverage. Short-fulcrum type has more closing force against needle valve.

CARBURETOR ANGULARITY

Visualize this problem as you tip the carburetor. Consider how the fuel depth (head) changes relative to the main jets, discharge nozzle and fuel-bowl vents when the carburetor is tipped. You might actually take a bowl with some fuel in it and observe its movement.

Note that the angle can affect the height of the discharge nozzle in relation to the fuel level. In some attitudes, a 40° tilt may cause fuel to drip or spill over from the discharge nozzle. This may require you to lower the fuel level. Increase the mechanical setting 1/32—1/16 in. per change. These changes are going to be more effective with a small inlet valve.

The problems with lowering the fuel level are: The power valve and main jets are uncovered sooner and nozzle "lag" may create a noticeable off-idle bog. Part of this lag can be offset with richer metering, which improves nozzle response. As with all other changes, make them in fine increments and don't go overboard.

Float Setting Change vs. Inlet Size

Fuel Pressure (psi)	Inlet Size (inch)	Fuel Level	Float Setting
6.0	0.110	0.600	0.160 ± 0.010 to toe
6.0	0.125	0.600	0.265 ± 0.010 to toe
6.0	0.135	0.600	0.296 ± 0.010 to toe

Relation of Inlet Size & Fuel Pressure to Float Setting

Carburetor		Fuel Pressure (psi)	Inlet Size (inch)	Fuel Level	Float Setting
7043244	Buick	4.5	.125	.750	.479 ± 010 to toe
7043243	Buick	5.25	.135	.655	.405 ± 010 to toe
7043515	Chevrolet	8.0	.125	.600	.355 ± 010 to toe
7043513	Chevrolet	6.0	.125	.600	.265 ± 010 to toe
7043274	Pontiac	6.0	.135	.675	.390 ± 010 to toe

Four thin filters with metal-mesh outside covering have been glued together with RTV. Hood forms base for cleaner, which is ducted to top of carburetor.

checks under as near the same conditions as possible. If your times are still good, decrease the size still more. Continue dropping the size and making tests until the accelerations fall off slightly.

Now go up to the next size larger inlet valve and that part is settled. It is likely that the inlet valve will end up smaller for driving off-road.

Vent Tubes—Vent tubes should be extended up as high as possible to handle sloshing fuel. It can come out of the vent tubes under severe braking and/or bouncing. Off-road cars with remotely mounted air cleaners are sometimes equipped with appropriately sized hoses to extend the vents all the way up into the air cleaner.

Don't extend them up against an air cleaner top and cause a restriction. Also remember to increase their size accordingly if you make the hoses exceptionally long. A restriction from length or obstruction could cause you real metering problems.

Air cleaner—The best possible air cleaner is one that completely encloses the carburetor, with the throttle operated by a cable. The cable housing can be sealed where it enters the baseplate for the air cleaner. More details on off-road air cleaners are in HP Books' *Off-Roader's Handbook* and *Baja Bugs & Buggies*.

Should you choose not to completely enclose the carburetor, be sure to inspect the cleaner exterior for any possible leak. Silt and microscopic dust particles will

Wrap-around polyfoam over paper filter is often used for extra filtration. This combination really works.

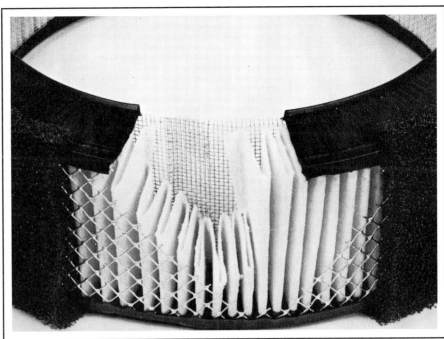

Dual-stage filter is almost unpluggable because foam outer layer keeps dirt from packing up inside folded paper. When dirt builds up, it falls off, so filter is somewhat self-cleaning. Highly recommended.

Centrifugal filters are popular with off-roaders (and contractors) because dust can be dumped easily to clean filter.

Oil-wetted cotton guaze filter is excellent for off-road racing. Photo by Spencer Murray.

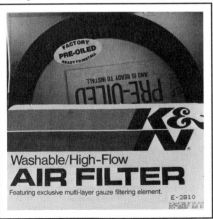

Vent (A) has been sealed with RTV compound. Choke assembly has been removed and screw holes and vacuum passage (B) sealed with same material.

shorten the engine's life, even if you overlook the smallest crack.

Silicone rubber is really great for plugging holes. It should be used freely. The final time you install the cleaner or air-duct adapter to the top of the carburetor, seal it all the way around the gasket. Better yet, don't use a gasket. It can get cocked or creep away from the flange.

Throttle Linkage—Use a cable throttle linkage for off-road installations to transmit less vibration to the driver's foot. Many stock vehicles have them, so fitting one to your application shouldn't be a problem. If it is light and small, carry an extra. Cables can bind because of fine dust.

Hydraulic linkages are also popular, but it is quite easy to overstress the throttle levers with such hookups. The hydraulic linkage should be installed so that the full travel of the actuating cylinder provides WOT and *no more*. In some instances, the cylinder mounting has to be carefully thought out so that there is not an over-center condition during throttle operation. For very light vehicles, you may decide the weight of these units is objectionable.

PERFORMANCE TUNING ELECTRONIC Q-JETS

First, read the sections "Electronic Carburetion" in chapter 4, Quadrajet Design and "Servicing Electronic Q-jets—E4ME/C" in chapter 5, Quadrajet Service. These will explain the basics of theory and service for the electronic Q-jet. This section is written with the assumption that you are familiar with that information.

By its very design, a microprocessor-controlled carburetor will try to maintain the ideal 14.7:1 A/F ratio for emissions and fuel economy reasons. There are certain times the ECM abandons the 14.7:1 ratio, and these can be used for performance gains.

The two times the ECM will abandon the ratio are during cold start/warmup and during heavy to WOT acceleration. My comments are about improving performance during heavy to WOT acceleration.

WOT Signals—The TPS tells the ECM how far the throttle blades are opened. If the TPS registers WOT, it sends a signal to the ECM, which then puts the carburetor into special operation for WOT.

Essentially, this case allows the carburetor to supply a richer mixture for best power. The carburetor is going from closed

Microprocessor-controlled Q-jets restrict high-performance gains average tuner can make. But, performance can be improved during heavy to WOT operation.

Expert tuners have full complement of metering rods, jets and hangers for dialing in performance on electronic Q-jets.

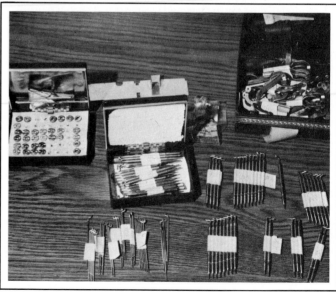

loop (emission calibration) to open loop (cold start/performance mixture). When the ECM is controlling WOT carburetor mixtures, it also controls timing and emission controls for performance.

What can you do to increase performance? Imagine the electronic carburetor's open-loop operation as equivalent to conventional metering. The TPS doesn't tell the ECM to keep the mixture at 14.7:1 from approximately 55°—60° throttle opening to WOT. So, most performance gains are limited to changing the secondary metering. Specifics for that are detailed earlier in this chapter, in "Drag Racing Modifications" and on page 149.

Of course, eventually you will come off WOT and the carburetor will regain the 14.7:1 air/fuel ratio.

Exhaust Changes—Any improvement in exhaust flow will result in a performance gain. A set of headers that retain the oxygen sensor and a lower-restriction muffler are a good start. For street applications, dual exhausts are often most effective. In many cases, you can also change to a better flowing catalytic converter.

Camshaft—The next step should be to install a slightly different cam. Exercise care when choosing a performance cam. If the profile is changed too much from stock, manifold vacuum will be altered so much that wrong signals are sent to the ECM.

Take a conservative approach. Increase camshaft lift first. It's when large changes are made in overlap, duration and so forth, that the ECM gets data it can't process.

Fuel Metering—Most microprocessor-controlled Q-jets are rated at 590—655 cfm (50°—70° air-valve opening)—adequate for street performance. Change the secondary metering rods to increase fuel delivery to the secondaries. Make sure the secondary air blades open fully. If they don't, shorten the air-valve control stop to allow full movement. This requires changing the vacuum break air valve damping diaphragm assembly for sufficient travel.

In some full-on performance applications you may want to change your primary jets. Use caution when doing this. Don't exceed the limits of the microprocessor, or it can't maintain a 14.7:1 A/F ratio under normal operating conditions. There is usually some adjustment tolerance with the primary jets, but be aware of the limitations imposed by the ECM.

Recheck the specifications for the M/C

solenoid after making metering changes. In most cases, it shouldn't require readjustment, but if it does, now is the time to do it.

Timing—Advance the basic engine timing when the previous modifications have been completed. If the car has a knock sensor, it can be used to your advantage when making adjustments. If you advance the timing too much, the sensor will tell the ECM to retard the timing and prevent engine damage.

If the engine doesn't have a knock sensor, make conservative adjustments. Increase the timing using the old tried-and-true way of advancing in two-degree increments until detonation (pinging) occurs, then retard the timing back two or more degrees. The ECM will continue to do its job with the timing advanced modest amounts. Don't increase the basic timing very far from the factory setting.

Racing Performance—If full race performance is your objective, the best *current* route to achieve it is return to an engine without microprocessor control of carburetion. Racing performance with microprocessor-controlled engines is achievable, but expensive.

DON'T FRY THE ECM

One special note: When connecting external gauges, driving lights, or accessories requiring electrical power, tap directly to the battery or ignition switch circuit. *DON'T* splice randomly into the wiring harness. If you happen to connect to a power source to or from the ECM, the resulting power surges or glitches can confuse or damage it. It is expensive to replace.

Edelbrock "Performer" manifold with "Performer Plus" camshaft and lifters for big block Pontiac. Photo courtesy Edelbrock.

Use quality timing light when advancing timing on ECM-controlled engines without knock sensors.

TPS signals ECM how far throttle blades are open from this connector.

8

Tuning Tips

These tips apply mainly to RPD carburetors that are *not microprocessor controlled*. Consequently, most of the discussion is limited to modifying Q-jets and two-barrels from 1965—80. Beginning in 1981, all GM passenger vehicles with RPD carburetors had microprocessor-controlled idle, off-idle and primary main metering systems. You are better off not altering these OEM carburetors to achieve better economy.

Suppose you have only one vehicle and a strong desire to do some weekend driving in motorsports events: slaloms, gymkanas, rallies, road races, drag races or hill climbs. You know a stock engine hasn't a prayer of being competitive in an off-road event where modifications are permitted. You also know that you must sacrifice some driveability and loss of operating economy on the street to be reasonably competitive in the off-road event of your choice. And, you are willing to accept these sacrifices—up to point. But where is that point?

That's a good question—and a difficult one, compounded by increasingly stringent emissions limits. You may be called upon at any time for a roadside emissions check by state or city authorities. Some states have laws requiring owners to return a vehicle to production specifications if it fails this test.

My advice is to keep all emissions-control devices hooked up and operative. These devices *do* help reduce exhaust emissions, and give evidence at any vehicle inspection that your intent is to conform with the law, even if the exhaust-emission levels are unacceptable.

Because most of you will be interested in getting more economy from a Q-jet, I'll begin with what might be done to improve its economy, driveability and general performance. Much of what we discuss for the Q-jet will apply to the two-barrels.

Thorough tuning requires patience, tools and time to experiment.

Q-JET MODIFICATIONS

Idle Systems—Do not alter the idle systems unless off-idle or light-throttle acceleration sags exist. Nearly any modification made here will affect the engine's ability to pass emissions inspections.

The channel-restriction orifices can be easily altered if you have a suitable set of small drills. The only reason to do this is if light-throttle weakness and surge exist. Increasing the diameter a few thousandths of an inch will richen the off-idle mixture. Never exceed a 10% increase in orifice area. Use tacky grease on the bit to collect tiny chips as you drill the thin orifice. A 10% increase in channel-restriction-orifice area is equivalent to approximately a 0.003-in. increase in orifice diameter.

Q-jet idle tubes are difficult to alter. Their diameter is 0.030—0.035 in. and standard drill bits that size are approximately 1-1/2-in. long. The tube orifice is 1-21/32 in. below the tube's top. The bit will not reach the calibrated orifice and enlarging the upper portion of the tubes does not help. Extra-long drills in the required size range are difficult to find. Idle tubes are literally impossible to remove without a special tool. And, these tools and parts are not dealer-service items, so don't expect help from your regular parts man or service manager. They do not have the parts or the required tools.

There are two reasons why you might want to increase the idle-tube orifice size. First, when curb-idle is not satisfactory with the idle-mixture needles turned out (counterclockwise), larger idle tubes will give more mixture control. Second, Q-jets are sometimes plagued with *nozzle drip* at idle, which causes poor idle stability.

This can be observed by taking off the air cleaner and looking into the carb while the engine is idling. Increasing idle-mixture richness with the adjusting screws will cure many cases of nozzle drip, but it sometimes takes larger idle orifices to do so. Larger orifices also help light-throttle driving during maneuvers from idle to 30 mph.

Should you enlarge the orifices, be sure to get all the chips out. Remove the primary jets and force fuel and air back and forth through the idle tube via the jet wells. When enlarging the orifices, do not exceed the stock diameter by more than 0.005 in.

Remember, any change in metering-orifice size will affect emissions and should only be done for off-road driving.

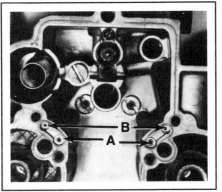

Idle tubes (A) have restrictions 1-21/32-in. below upper end. Long drill or special reamer is required to enlarge them without removal. Channel restrictions (B) can be opened to improve mid-range response.

Tapered reamer being used to open channel restriction orifice just enough to cure mid-range driving problems.

Special drill being inserted to enlarge idle-tube metering restriction. This modification means engine won't pass emission tests at speeds above curb-idle.

Secondary Metering—Once you try metering the Q-jet to meet specific standards, it becomes an enjoyable challenge. This carburetor has many primary rods, secondary rods and hangers available for making infinite variations in the A/F ratio to meet engine demands.

Suppose you want to improve low-speed driveability and economy, but you also want to be able to run the engine near WOT for extended periods of time. Because the primary metering cannot be changed without losing economy and driveability during normal driving, the extra fuel must come from the secondary metering system.

The secondary metering jets are a fixed size and not alterable, but at this writing there are 93 different rod sizes/shapes available with various tapers and tip diameters. For extremely fine tuning at the factory, there are also 20 different rod-hanger heights. However, only one is available from Delco as a replacement part.

If you have tuned the off-idle and primary metering systems to run at WOT for extended periods and the carb is now leaner than stock, additional fuel must come from the secondaries.

CHART A
METERING-AREA QUICK REFERENCE CHART

With this chart you can read the calculated area determined for each of the listed diameters.

$$\pi \left(\frac{d}{2}\right)^2 = A$$

$$d^2 \times \frac{\pi}{4} = A$$

$$d^2 \times 0.7854 = A$$

INCH	MM	SQ. IN.	DRILL SIZE
.001	.025	.000000785	
.002	.051	.000003142	
.003	.076	.000007069	
.004	.102	.000012566	
.005	**.127**	**.000019635**	
.006	.152	.000028274	
.007	.178	.000038484	
.008	.203	.000050265	
.009	.229	.000063617	
.010	**.254**	**.000078540**	
.011	.279	.000095033	
.012	.305	.000113097	
.013	.330	.000132732	80 (.0135)
.014	.356	.000153938	79 (.0145)
.015	**.381**	**.000176714**	78 – 1/64 (.0156)
.016	.406	.000201062	
.017	.432	.000226980	
.018	.457	.000254469	77
.019	.483	.000283528	
.020	**.508**	**.000314159**	76
.021	.533	.000346360	75
.022	.559	.000380132	74 (.0225)
.023	.584	.000415475	
.024	.610	.000452389	73
.025	**.635**	**.000490873**	72
.026	.660	.000530929	71
.027	.686	.000572555	
.028	.711	.000615752	70
.029	.737	.000660519	69 (.02925)
.030	**.762**	**.000706858**	
.031	.787	.000754767	68–1/32 (.0312)
.032	.813	.000804247	67
.033	.838	.000855298	66
.034	.864	.000907919	
.035	**.889**	**.000962112**	65
.036	.914	.001017875	64
.037	.940	.001075209	63
.038	.965	.001134114	62
.039	.991	.001194590	61
.040	**1.016**	**.001256636**	60
.041	1.041	.001320253	59
.042	1.067	.001385441	58
.043	1.092	.001452200	57
.044	1.118	.001520530	
.045	**1.143**	**.001590430**	
.046	1.168	.001661901	
.047	1.194	.001734943	56 (.0465)–3/64 (.0468)
.048	1.219	.001809556	
.049	1.245	.001885739	
.050	**1.270**	**.001963494**	
.051	1.295	.002042819	
.052	1.321	.002123715	55
.053	1.346	.002206182	
.054	1.372	.002290219	
.055	**1.397**	**.002375827**	54
.056	1.422	.002463007	
.057	1.448	.002551756	
.058	1.473	.002642077	
.059	1.499	.002733969	53 (.0595)
.060	**1.524**	**.002827431**	
.061	1.549	.002922464	
.062	1.575	.003019068	1/16 (.0625)
.063	1.600	.003117243	52 (.0635)
.064	1.626	.003216988	
.065	**1.651**	**.003318304**	
.066	1.676	.003421192	
.067	1.702	.003525649	51
.068	1.727	.003631678	
.069	1.753	.003739277	
.070	**1.778**	**.003848448**	50

INCH	MM	SQ. IN.	DRILL SIZE
.071	1.803	.003959189	
.072	1.829	.004071501	
.073	1.854	.004185383	49
.074	1.880	.004300837	
.075	**1.905**	**.004417861**	
.076	1.930	.004536456	48
.077	1.956	.004656622	
.078	1.981	.004778358	5/64 (.0781)–47 (.0785)
.079	2.007	.004901666	
.080	**2.032**	**.005026544**	
.081	2.057	.005152993	46
.082	2.083	.005281013	45
.083	2.108	.005410603	
.084	2.134	.005541765	
.085	**2.159**	**.005674497**	
.086	2.184	.005808800	44
.087	2.210	.005944673	
.088	2.235	.006082118	
.089	2.261	.006221133	43
.090	**2.286**	**.006361720**	
.091	2.311	.006503876	
.092	2.337	.006647604	
.093	2.362	.006792903	
.094	2.388	.006939772	
.095	**2.413**	**.007088212**	42 (.0935)–3/32 (.0937)
.096	2.438	.007238223	41
.097	2.464	.007389805	
.098	2.489	.007542957	40
.099	2.515	.007697681	
.100	**2.540**	**.007853975**	39 (.0995)
.101	2.565	.008011840	
.102	2.591	.008171275	38 (.1015)
.103	2.616	.008332282	
.104	2.642	.008494859	37
.105	**2.667**	**.008659007**	
.106	2.692	.008824726	
.107	2.718	.008992015	36 (.1065)
.108	2.743	.009160876	
.109	2.769	.009331307	
.110	**2.794**	**.009503309**	7/64 (.1094)
.111	2.819	.009676882	35
.112	2.845	.009852026	34
.113	2.870	.010028740	33
.114	2.896	.010207025	
.115	**2.921**	**.010386881**	
.116	2.946	.010568308	32
.117	2.972	.010751306	
.118	2.997	.010935874	
.119	3.023	.011122013	
.120	**3.048**	**.011309723**	31
.121	3.073	.011499004	
.122	3.099	.011689856	
.123	3.124	.011882278	
.124	3.150	.012076271	
.125	**3.175**	**.012271836**	1/8
.126	3.200	.012468971	
.127	3.226	.012667676	
.128	3.251	.012867953	30 (.1285)
.129	3.277	.013069800	
.130	**3.302**	**.013273218**	
.131	3.327	.013478206	
.132	3.353	.013684766	
.133	3.378	.013892896	
.134	3.404	.014102597	
.135	**3.429**	**.014313869**	
.136	3.454	.014526712	29
.137	3.480	.014741125	
.138	3.505	.014957110	
.139	3.531	.015174665	
.140	**3.556**	**.015393791**	28 (.1405)–9/64 (.1406)

Secondary metering performance improvements of microprocessor-controlled Q-jets are as easy as removing screw (arrow), lifting out hanger and changing rods.

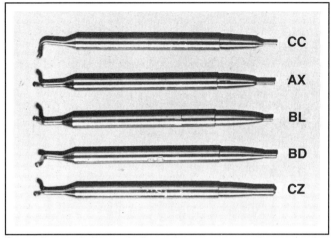

Secondary metering rods tips and taper vary as mixture requirements do. Super-rich rod at top is CC, followed by AX, BL, BD and CZ (super-lean).

Q-JET MAIN METERING JETS

Q-jet primary jets differ from those in other RPD models. *Don't* interchange them. They are identified by two curved lines stamped opposite identification number on jet face. This number specifies orifice size in thousandths of an inch.

Part No.	Stamped On Jet	Diameter Of Orifice
7031970	70	.070 in.

PRIMARY MAIN METERING JETS

Part No.	Stamped On Jet
7031966	66
7031967	67
7031968	68
7031969	69
7031970	70
7031971	71
7031972	72
7031973	73
7031974	74
7031975	75
7031976	76
7031977	77
7031978	78

Q-JET PRIMARY MAIN METERING RODS

Two primary metering rod types are used. Some have a single taper at metering tip; others have a double taper at tip. Both types use a similar two-digit numbering notation.

The number specifies metering rod diameter at point A and is the last two digits of the part number. Rods with a double taper have B stamped on the rod after the number.

All rods have a 0.026-in.-diameter tip at their smallest point, except rod 50D, which has a 0.036-in. tip.

Note: Q-jets beginning with serial number 170 have primary metering rods that are 0.080-in. *shorter* and with a different taper. These started in 1975 and continue to the present. The different taper and lengths mean these later primary rods are *not* interchangeable with previous (pre-'75) ones.

Remember, *do not* interchange 170- - - - - and 70- - - - - metering rods. 70- - - - - rods and carburetors *were* used in many applications thru the mid-'80s.

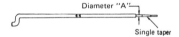

1965—67 RODS

Part No.	Stamped No.	Diameter at A
7031844	44	0.044 in.

PRIMARY MAIN METERING RODS
(Single Taper)

Part No.	Stamped
7031833	33
7031835	35
7031836	36
7031837	37
7031838	38
7031839	39
7031840	40
7031841	41
7031842	42
7031843	43
7031844	44
7031845	45
7031846	46
7031847	47
7031849	49

1968 RODS

Part No.	Stamped No.	Diameter at A
7034844	44B	0.044 in.

PRIMARY MAIN METERING RODS
(Double taper)

Part No.	Stamped
7034836	36B
7034838	38B
7034839	39B
7034840	40B
7034841	41B
7034842	42B
7034843	43B
7034844	44B
7034845	45B
7034846	46B
7034847	47B
7034848	48B
7034849	49B
7034850	50B
7034851	51B
7040699	48C*
7040701	52C*
7046388	50D**

* Special triple-taper rods—used 1970 Oldsmobile.

** Special rod used in GMC Motorhome—has 0.036-in. power tip.

All secondary metering orifices are fixed at 0.135-in. diameter so the area will be constant at 0.0143139 sq. in. At WOT, approximately 60% of the total mixture flow through the carburetor is supplied by the secondaries. For this reason, a given percentage change in secondary metering would have half again as much effect on the overall A/F ratio as an identical percentage change in primary metering. The secondary flow area is approximately 1.5 times the primary area. This factor must be considered when changing secondary metering. Seldom will your calculations exactly match an available rod, so an intelligent compromise is almost always required.

If no replacement secondary metering rods of the desired type are available, you can reduce the size of the power tip on the rods you have. Carefully chuck each rod in a lathe or drill press and use emery cloth or a pattern file to reduce the diameter. Frequently check with a micrometer or caliper to make sure you are not removing too much material.

This method should be regarded as a temporary measure and only suitable for special performance applications. Removing the slightly hardened surface of a metering rod causes it to wear excessively and very quickly.

Power Mixture—Manifold vacuum provides the signal to control power-system fuel flow. Some late-model cars have low numerical differential-gear ratios and mild camshafts. These cause manifold vacuum

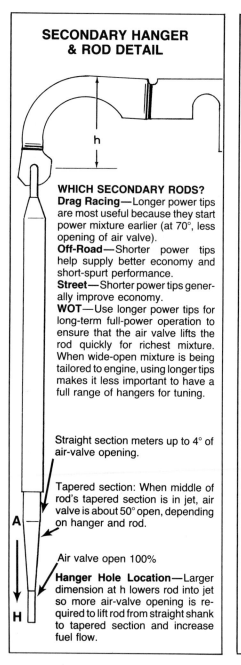

SECONDARY HANGER & ROD DETAIL

WHICH SECONDARY RODS?
Drag Racing—Longer power tips are most useful because they start power mixture earlier (at 70°, less opening of air valve).
Off-Road—Shorter power tips help supply better economy and short-spurt performance.
Street—Shorter power tips generally improve economy.
WOT—Use longer power tips for long-term full-power operation to ensure that the air valve lifts the rod quickly for richest mixture. When wide-open mixture is being tailored to engine, using longer tips makes it less important to have a full range of hangers for tuning.

Straight section meters up to 4° of air-valve opening.

Tapered section: When middle of rod's tapered section is in jet, air valve is about 50° open, depending on hanger and rod.

Air valve open 100%

Hanger Hole Location—Larger dimension at h lowers rod into jet so more air-valve opening is required to lift rod from straight shank to tapered section and increase fuel flow.

Part Number	I.D. Code	A	B	C	D	E	F	G	H	Tip Lgth
7033772	AD	.1346	.1280	.1084	.0812	.0627	.0397	.0222	—-	—
7042335	CB	.1347	.1287	.1197	.0197	.0897	.0547	.0297	.0279	S
7040724	BV	.1347	.1287	.1197	.1097	.0897	.0547	.0297	.0297	S
17064719	DX	.1320	.1247	.1090	.0600	.0400	.0300	.0300	.0300	M
7047816	DC	.1347	.1236	.1129	.1024	.0599	.0300	.0300	.0300	M
7042356	CC	.1347	.1236	.1129	.1024	.0599	.0300	.0300	.0300	M
7040856	BY	.1287	.1217	.1062	.0817	.0667	.0317	.0317	.0317	M
17059952	DU	.1280	.1244	.1075	.0847	.0547	.0337	.0337	.0337	S
17056618	DS	.1310	.1244	.1075	.0847	.0547	.0337	.0337	.0337	M
7048890	DG	.1342	.1244	.1075	.0847	.0547	.0337	.0337	.0337	M
7048512	DF	.1339	.1262	.1075	.0847	.0547	.0337	.0337	.0337	M
7044775	CF	.1347	.1244	.1075	.0847	.0547	.0337	.0337	.0337	M
17050221	DM	.1334	.1281	.1155	.0853	.0655	.0482	.0393	.0393	S
7038034	BP	.1337	.1287	.1197	.1097	.0397	.0687	.0397	.0397	S
7040767	BW	.1317	.1247	.1087	.0897	.0597	.0397	.0397	.0397	M
7045780	CJ	.1339	.1244	.1075	.0807	.0419	.0419	.0397	.0397	S
7037744	BM	.1107	.1290	.1159	.0932	.0519	.0397	.0397	.0397	M
7034822	BG	.1352	.1294	.1165	.0932	.0519	.0397	.0397	.0397	M
7035916	BH	.1342	.1290	.1159	.0932	.0519	.0397	.0397	.0397	M
7033549	AX	.1339	.1244	.1075	.0807	.0419	.0419	.0397	.0397	L
7034335	BB	.1339	.1244	.1075	.0807	.0419	.0419	.0397	.0397	L
7034400	BF	.1325	.1315	.1192	.1021	.0891	.0639	.0397	.0397	S
7036077	BJ	.1333	.1315	.1193	.1021	.0891	.0639	.0397	.0397	S
7037295	BK	.1325	.1315	.1193	.1021	.0891	.0639	.0397	.0397	S
7042304	CA	.1322	.1277	.1147	.0937	.0797	.0397	.0397	.0397	M
7045924	CS	.1330	.1287	.1197	.1097	.0897	.0687	.0397	.0397	S
7045840	CM	.1342	.1294	.1165	.0932	.0519	.0397	.0397	.0397	M
7043771	CE	.1347	.1244	.1075	.0847	.0547	.0410	.0410	.0410	M
7036671	BN	.1329	.1288	.1147	.0957	.0793	.0503	.0410	.0410	S
7037733	BL	.1329	.1311	.1188	.1021	.0891	.0691	.0410	.0410	S
7034377	BE	.1329	.1311	.1188	.1021	.0891	.0691	.0410	.0410	S
7046010	DA	.1333	.1244	.1075	.0847	.0577	.0440	.0440	.0440	M
7046004	CY	.1331	.1244	.1075	.0847	.0577	.0440	.0440	.0440	M
7033655	AU	.1345	.1291	.1154	.0927	.0527	.0527	.0527	.0527	L
7045781	CK	.1345	.1291	.1154	.0927	.0527	.0527	.0527	.0527	L
7045984	CV	.1332	.1291	.1154	.0927	.0527	.0527	.0527	.0527	L
7033812	AH	.1345	.1291	.1154	.0927	.0727	.0527	.0527	.0527	M
17081930	EJ	.1355	.1350	.1200	.1047	.0837	.0687	.0540	.0540	S
7040725	BU	.1347	.1262	.1172	.1097	.0997	.0747	.0547	.0547	S
17080091	EA	.1339	.1262	.1172	.1000	.0825	.0721	.0547	.0547	S
7045923	CR	.1339	.1262	.1172	.1097	.0997	.0747	.0547	.0547	S
7045985	CX	.1346	.1294	.1165	.0932	.0567	.0567	.0567	.0567	L
17082409	EK	.1340	.1294	.1165	.0932	.0790	.0567	.0567	.0567	S
17081878	ED	.1332	.1247	.1090	.0879	.0743	.0578	.0567	.0567	M
17081880	EF	.1340	.1294	.1165	.0932	.0790	.0617	.0567	.0567	S
17081881	EG	.1350	.1292	.1157	.0957	.0847	.0697	.0567	.0567	S
17075268	DZ	.1325	.1290	.1150	.0879	.0743	.0578	.0567	.0567	S
7048992	DH	.1332	.1247	.1090	.0879	.0743	.0578	.0567	.0567	S
7045842	CP	.1320	.1247	.1090	.0879	.0743	.0578	.0567	.0567	M

Q-JET SECONDARY METERING RODS

to fall quickly whenever the throttle is opened. The stock power systems feed fuel on medium-throttle, light accelerations when the added fuel is not really required. Economy suffers.

For economical operation it's a good idea to change the power-system cut-in point to a lower vacuum setting so the driver will not be "into the power system" so much of the time during normal driving. Changing the power-spring length changes the cut-in point. By making this simple change, greater fuel economy results.

As with nearly any improvement, there are compromises. Reduction of power-system operating range may cause leanness during medium-throttle accelerations. This will seldom be objectionable, except perhaps during engine warmup following cold starts—typically for the first mile or so after the choke comes off.

Q-jet power spring No. 7037851 has the ideal 3—6-in.Hg vacuum operating characteristics. You can use this standard part instead of cutting coils off the existing spring in your Q-jet. If you want to cut off coils, start with 1-1/2 coils, counting the end coil as a full coil. Two full coils is the absolute maximum to cut off a standard Q-jet power-piston spring.

After the spring has been cut, the piston must still return fully and decisively to the top of its bore when you push it down and release it. If it doesn't, too much has been cut from the spring and the power system will not operate correctly.

It's a good idea to check the action of the power piston in its bore before disassembling the carburetor so you will have a baseline to compare with. Measure the height of the primary-metering-rod hanger above the gasket surface with the stock spring installed. Use this measurement as a guide to check against when you install a modified spring.

If you are driving a car with a high numerical differential-gear ratio, you may not need to change the power-system spring because more throttle can be used without causing drastic reductions in manifold vacuum. This is especially true with mild cam timing. Careful observation of manifold vacuum during accelerations can help you decide when you want the power system to operate.

Camshaft—Modern engines use a camshaft designed for emissions reduction when combined with low numerical axle ratios. This combination restricts the engine to a rpm range where it can never get much manifold vacuum. If the gear ratio were to be swapped for a higher numerical ratio (lower gear), the camshaft would probably work better because the engine might be able to take some advantage of the overlap characteristics.

If you are seriously interested in performance and keeping some economy, replace the stock camshaft with one that is more efficient for modest speed operation and today's low-compression engines. Many camshaft grinders offer camshafts specifically designed for emissions-controlled engines. Installing one will generally increase performance without much sacrifice in economy.

Summary—Make metering changes in progressive steps. The primary main metering system should never be calibrated too lean. A modest leaning benefits economy. The reasoning is that a modestly lean part-throttle is more responsive and the

Part Number	I.D. Code	A	B	C	D	E	F	G	H	Tip Lgth
7045341	CN	.1340	.1294	.1165	.0932	.0790	.0617	.0567	.0567	S
7045779	CH	.1350	.1292	.1157	.0957	.0847	.0697	.0567	.0567	S
7042719	CD	.1351*	.1264	.1107	.0755	.0567	.0567	.0567	.0567	L
7042300	BZ	.1352*	.1294	.1165	.0932	.0567	.0567	.0567	.0567	L
7034337	BA	.1332	.1247	.1090	.0879	.0743	.0578	.0567	.0567	M
7033981	AP	.1345*	.1292	.1132	.0907	.0757	.0567	.0567	.0567	M
7033889	AZ	.1350	.1292	.1157	.0957	.0567	.0567	.0567	.0567	L
7033830	AY	.1352	.1294	.1165	.0932	.0567	.0567	.0567	.0567	L
7033680	AL	.1349	.1247	.1090	.0879	.0743	.0578	.0567	.0567	M
7033628	AJ	.1345	.1292	.1132	.0907	.0757	.0567	.0567	.0567	M
7033182	AV	.1338	.1292	.1132	.0907	.0757	.0567	.0567	.0567	M
7033171	AR	.1352*	.1294	.1165	.0932	.0790	.0617	.0567	.0567	M
7033104	AK	.1350	.1292	.1157	.0957	.0847	.0697	.0567	.0567	S
17081879	EE	.1323	.1250	.1093	.0882	.0746	.0581	.0570	.0570	M
17062170	DV	.1348	.1294	.1168	.0935	.0793	.0620	.0570	.0570	S
17053659	DR	.1323	.1250	.1093	.0882	.0746	.0581	.0570	.0570	M
7034365	BD	.1332	.1312	.1168	.0939	.0805	.0577	.0577	.0577	M
7034300	BC	.1352*	.1294	.1174	.1015	.0901	.0599	.0581	.0581	S
17053703	DN	.1355	.1297	.1108	.1019	.0904	.0602	.0584	.0584	S
7040601	BT	.1338	.1292	.1132	.0907	.0787	.0597	.0597	.0597	M
17081095	EC	.1350	.1292	.1157	.0847	.0697	.0610	.0610	.0610	M
7048908	DJ	.1337	.1262	.1172	.1097	.0997	.0847	.0747	.0647	S
7045782	CL	.1350	.1294	.1176	.0984	.0667	.0667	.0667	.0667	L
7033658	AT	.1350	.1294	.1176	.0984	.0667	.0667	.0667	.0667	L
7048892	DL	.1336	.1260	.1127	.0939	.0824	.0711	.0679	.0679	S
17081882	EH	.1325	.1267	.1137	.0945	.0829	.0721	.0686	.0686	S
17053531	DP	.1325	.1267	.1137	.0945	.0829	.0721	.0686	.0686	S
7040797	BX	.1352*	.1294	.1177	.1010	.0907	.0797	.0697	.0697	S
7034320	AN	.1352*	.1294	.1177	.1010	.0907	.0797	.0697	.0697	S
7047806	DB	.1334	.1294	.1177	.1010	.0907	.0797	.0697	.0697	S
17056651	DT	.1337	.1265	.1180	.1100	.1000	.0700	.0700	.0700	M
7045983	CT	.1337	.1294	.1176	.0984	.0849	.0774	.0774	.0774	M
7038256	AS	.1350	.1294	.1176	.0984	.0849	.0774	.0774	.0774	M
7045778	CG	.1350	.1294	.1176	.0984	.0849	.0774	.0774	.0774	M
17074266	DY	.1337	.1305	.1191	.1010	.0883	.0874	.0874	.0874	L
17062432	DW	.1334	.1294	.1170	.1010	.0883	.0874	.0874	.0874	L
7048092	DE	.1337	.1305	.1191	.1010	.0883	.0874	.0874	.0874	M
7038910	BR	.1350	.1294	.1176	.0984	.0897	.0897	.0897	.0897	L
7033194	AW	.1351*	.1283	.1209	.1036	.0965	.0965	.0905	.0905	M
7038911	BS	.1350	.1292	.1157	.0947	.0947	.0947	.0947	.0947	L
7045986	CZ	.1345	.1292	.1157	.0947	.0947	.0947	.0947	.0947	L
17080467	EB	.1345	.1292	.1157	.0947	.0947	.0947	.0947	.0947	L
7048919	DK	.1345	.1298	.1169	.0997	.0997	.0997	.0997	.0997	L
7048091	DD	.1345	.1298	.1169	.1047	.1047	.1047	.1047	.1047	L
		20°	40°	60°	70°	80°	90°	100°		
		(Degrees of air-valve opening)								

Note: Rods with part numbers beginning with 170 and 70 are interchangeable. ID code letters are stamped in rod. Asterisk after dimension A indicates slotted rod. Dimension H is smallest diameter and richest part of metering rod. Tip Length L supplies rich metering from 70°—100° of air-valve opening. S tips are from 90°—100°.

carb won't be into the WOT mixtures nearly as often. When you do "step on it" for passing and such maneuvers, the response will be crisp.

Secondary metering rods should be changed to provide added fuel required for extended WOT operation. This eliminates the possibility of frying pistons, valves and plugs on a long pull. Actually, you can skip this step if the car is not going to be used for anything other than general transportation. However, it is a good safety measure to protect the engine in the event someone else should use the car and not know of the carburetion changes.

If your main concern is WOT drag or road racing, the approach can be a bit different. Meter the primary WOT and secondary WOT fuel requirements first. If this ends up causing a part-throttle richness, increase the part-throttle metering-rod area to reduce part-throttle fuel flow. You might also need to install secondary metering rods with short power-tip ends so the intermediate opening positions are somewhat lean. With this metering you will undoubtedly compromise economy and affect emissions levels that are acceptable for street driving.

TWO-BARREL MODIFICATIONS

The modifications discussed here have been accomplished on two-barrel carburetors from Chevrolets, but the details also apply to Buicks, Oldsmobiles and Pontiacs.

You can usually improve economy and drivability by increasing the original main-jet size by one to two thousandths of an inch. This would mean replacing the size 50 jets with size 51s. Or, if the car is being driven in the extreme north where there is a lot of cold weather—or being used to pull a trailer or carry heavy loads—a size 52 might be even better.

Whenever a 2G carburetor has pullover tubes with small feed orifices above the venturi, you can nearly always improve highway cruise economy by plugging the tubes and increasing jet size by one or two thousandths of an inch.

When jet sizes are increased, the total fuel available at the nozzle is increased too. Therefore, it is wise to reduce the power restriction area approximately 5%. Removing the power restrictions so they can be replaced with smaller ones is somewhat difficult on late models with the Adjustable Part Throttle (APT) feature. This is be-

cause one of the plugs is larger and hard to remove. The two small plugs used on earlier models without APT are easy to get out. The nearby photos and captions detail how to accomplish this modification.

When installing larger main jets, it is important to plug the APT feed hole. The APT was factory-adjusted by a computer to supply a certain part-throttle A/F mixture with that particular carburetor. You are not adding as much fuel with the larger main jets because you have eliminated that part of the main-system fuel previously supplied by the APT.

Plugging the APT system results in better metering consistency. The new combination will give better transition operation (off-idle to start of main-system fuel flow) and better tip-ins while cruising (light accelerations). Although dynamometer tests will not show any power difference between a carburetor with or without APT, driving the vehicle clearly shows better nozzle response (hence the better tip-ins and transitions) *without* APT. This is because all the fuel comes directly through the main jets, with no obstacle courses for the fuel to traverse before getting to the main well.

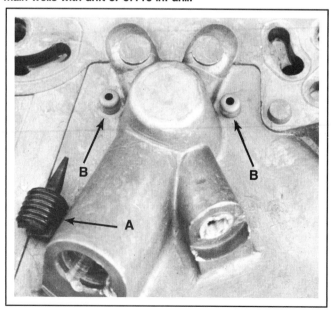

APT adjusting needle (A) can be eliminated. Cup plugs (B) are power-valve-channel restrictions that are driven through into main wells with drift or 0.110-in. drill.

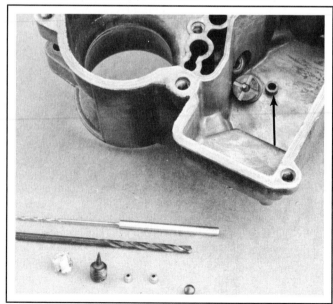

Top view shows APT feed hole (arrow) to be plugged with lead ball. Drill in holder is used to open idle-tube restrictions. Power restrictions must be replaced with smaller ones or soldered shut and redrilled for about 5% less area. When APT system is plugged, main jets can be increased 0.001 or 0.002 in.

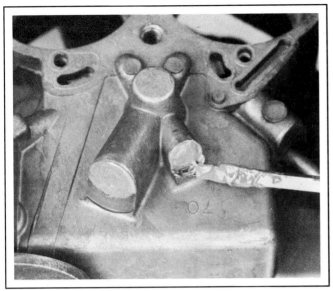
Epoxy drilled plugs in place.

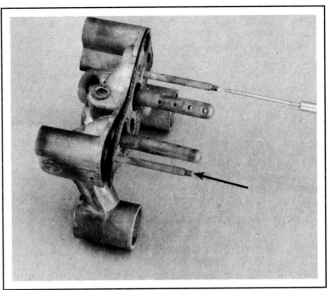
Cure tip-in sags by opening idle-tube orifices 0.002—0.003 in. for about 5% increase in area. This modification makes it impossible for engine to pass emissions tests at any speed above curb-idle.

Sandpaper wrapped around file is used to eliminate casting flash in venturis. Smoothing improves air-flow.

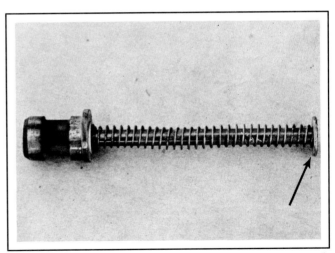
Three coils were cut from end of power-valve spring (arrow). Power piston and bore in air horn were polished with fine crocus cloth to improve piston action.

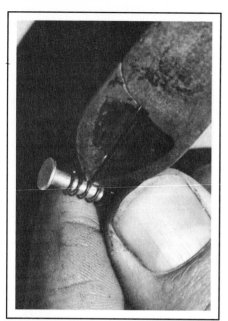

Cutting three full coils from 2G power-piston spring. Spring has been pulled up stem to allow easy coil counting.

In many instances, you can also remove approximately three coils from the power-piston assembly spring so more throttle can be used without using power-system fuel. See the preceding Q-jet "Power Mixture" section for more information on why this is important and how this modification will help economy.

If reducing the power-system fuel-flow area by 5% for economy, also reduce the power-valve channel-restriction area an additional 1% for each 1,000 feet you drive above sea level.

For example, if you live in Denver at 5,000 feet altitude, reduce the power-system fuel-flow area by 5% for the economy reduction, plus 5% for altitude, to make a 10% total reduction. If the car seems sluggish during passing maneuvers, increase the the area of power-system restrictions in 1% increments until you have obtained the best performance. In most cases, 10% leaner main system operation will improve driveability in all ranges.

If you are driving a car with an automatic transmission, a too-lean power mixture will tend to make the engine hover at the rpm just before the 2-3 shift. It is a good idea to note how crisply the shift occurs before making any carburetor modifications, preferably by timing several runs that include this shift point. Before making changes, be sure to run a baseline measurement as previously described. Then make similar tests after each area change.

If you plan to increase the main jet by two sizes, reduce the power-system restrictions and change the power-valve or power-system cut-in point *before* changing the main jets. You may find the stock jets to be correct once you have made the other changes and plugged the APT fuel supply.

Staking power-piston assembly in place by pushing part of metal over edge of retainer requires hammer and screwdriver.

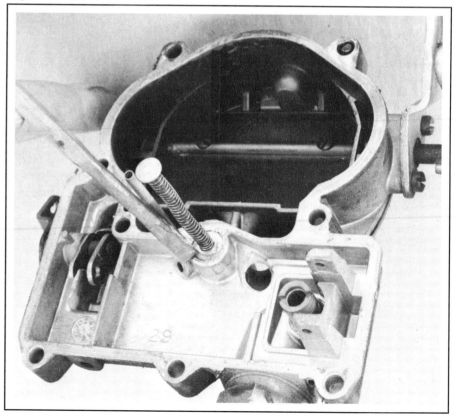

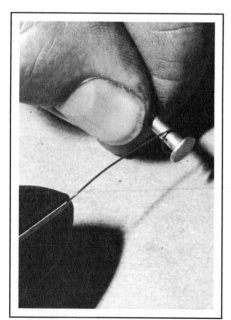

Once spring is cut, remaining spring is pulled out of way with fingers and cutters are used to unwrap cut-off coils from piston stem.

Emissions Control

9

TYPES OF EMISSIONS

Air pollution is a worldwide problem; it is not limited to urban or industrial areas. The magnitude of the problem and public concern are manifested in state and federal laws regulating type and amount of contaminants put into the atmosphere by automobiles, motorcycles, and light trucks. Standards for industrial emissions have also been set, but the major focus and enforcement has been directed at new automobile manufacturers and makers of after-market (replacement) parts that directly affect emissions. California, specifically the Los Angeles and San Francisco areas, has the greatest smog problem, and this state has led the others in emissions control regulations.

SMOG

What is *smog?* It's a simple term for a complex atmospheric condition. When unburned hydrocarbons (HC) and oxides of nitrogen (NOx) combine in the atmosphere and are acted upon by sunlight, chemical reactions produce *photochemical smog.* Their combination is aided by an inversion layer—a dense layer of air that prevents the escape of the ingredients into the upper atmosphere.

The inversion layer is often likened to a lid on a pot; it holds the ingredients so that the photochemical reaction has plenty of time to take place, often over several days.

Smog causes nose, throat and eye irritations. And, like carbon monoxide (CO), it is harmful to animal and plant life. Smog also deteriorates some plastics, paint, and

Carburetors to meet '80s emissions standards are tested in RPD's flow test area.

the rubber in tires, seals, weatherstripping and wiper blades.

HC, CO and NOx

Theoretically, perfect combustion of gasoline and oxygen would produce only carbon dioxide and water as emissions. Unfortunately, the pressure and temperature of the combustion chamber cause several undesirable combustion products. Two primary ones are HC and CO. They are partially burned A/F mixtures. Partial combustion is caused by a shortage of oxygen in the mixture. Combustion-chamber shape, cam overlap and other factors contribute to cylinder efficiency and these products.

CO forms whenever there is insufficient oxygen to complete the combustion process. Generally speaking, the richer the mixture, the higher the CO concentration. Even if the mixture is chemically correct, CO levels can't be reduced to zero because perfect mixing and cylinder-to-cylinder distribution are nearly impossible to achieve in carbureted engines.

Gasoline is composed of varied hydrogen and carbon compounds, hence the name *hydrocarbons*. Unburned hydrocarbons are just that—gasoline that did not get burned on its trip through the engine. There are several reasons why this happens. Rich mixture is one. Fuel that does not get burned because of misfiring is another. So, either a lean or rich A/F mixture can cause an increase in HC emissions. Other factors

affecting HC concentration include: high compression, spark advance, and high vacuum under deceleration.

Starting in 1966, California applied standards for tailpipe emissions of CO and HC, together with test procedures and sampling methods. The Federal government adopted the California test procedures and standards in 1968 and required, as California had earlier, all manufacturers to certify and prove their vehicles met the prescribed standards.

A third product emitting from combustion chambers is *oxides of nitrogen* (NOx). These form when the normally inert nitrogen present in the A/F mixture combines with oxygen under high temperatures and pressure. Lean mixtures that minimize CO and HC maximize NOx.

Reducing peak combustion temperature is one way to reduce NOx formation. This can be achieved by retarding the spark or diluting the incoming mixture. For many years, and to this date, the most practical and widespread method for this has been recirculating a portion of the exhaust gas. This is called *Exhaust Gas Recirculation (EGR)*.

In 1970, legislation was passed that started tests for checking HC, CO and NOx. Once test equipment became available and requirements and procedures were spelled out, manufacturers invented, designed, and tested many devices to meet standards. Let's examine these schedules.

EMISSIONS-TEST PROCEDURES

EXHAUST-GAS ANALYSIS

The early test procedure consisted of seven duplicate cycles with varying driving modes. A tester drove the vehicle on chassis rolls as directed by audible signals and visual instruments. Exhaust samples were taken at certain intervals during every mode. A computer then analyzed each mode individually and compounded the total at completion.

Separate modes were often run for expediency in testing. The numbers representing HC, CO, and NOx in a certified test were always taken from a full seven-cycle run following a 12-hour minimum wait in a 75F (24C) room. The first two cycles encompassed emissions during choke operation. By the end of the second cycle the automatic choke was off and the remaining cycles represented general operating conditions. The fifth cycle was run per schedule but not sampled.

Starting in 1972, actual dynamometer tests became part of a more complicated procedure. The seven-cycle test was replaced by a 23-cycle one that was considered by the automotive industry to be more representative of actual customer driving habits. Basically, the test consists of cold-start, idle-acceleration, steady-state deceleration, and idle conditions. This

Passenger Car Clean Air Act of 1970 Emission Control Requirements

	1971	1972	1973-4	1975-6	1977	1978-9	1980	1981-2	1983-6
HC (gm/mi) (California)	4.6	3.4	3.4	1.5 (.9)	1.5 (.41)	1.5 (.41)	.41	.41	.41
CO (gm/mi) (California)	47	39	39	15 (9.0)	15 (9.0)	15 (9.0)	7.0 (9.0)	3.4 (7.0)	3.4 (7.0)
NOx (gm/mi) (California)	- (4.0)	- (3.0)	3.0 (3.0)	3.1 (2.0)	2.0 (1.5)	2.0 (1.5)	2.0 (1.0)	1.0 (0.7)	1.0 (0.7)
SHED (gm/mi) (California)	- (6.0)	6.0#	2.0#	2.0#	2.0#	6.0*	6.0* (2.0)	2.0*	2.0*

Trapped Method
* SHED Method 1978 and thereafter

Fuel Economy Requirements

	1978	1979	1980	1981	1982	1983	1984	1985
MPG	18	19	20	22	24	26	27	27.5

Early '70s emissions testing on dynamometer to verify compliance during seven cycles (idle, accelerating and decelerating) of vehicle operation.

sequence is interrupted by a 10-min. soak period at cycle 18.

First, the vehicle is examined to ensure it is a production vehicle set to specifications. It then sits for at least 12 hours in a controlled atmosphere of 75F. Next, it is pushed onto the test dynamometer and the exhaust is attached to the emission sensors. This procedure is required to make sure full choke operation is included in the results. The exhaust gas is collected in a bag or chamber and is sampled continuously during all cycles. Computers and chart recorders monitor and register the HC, CO and NOx concentrations, and fuel economy.

The results of this test, conducted by the car manufacturer, and selectively verified by the Environmental Protection Agency (EPA) in their laboratories, determine whether the vehicle is emissions-certified for production.

SHED TESTS

Besides the emissions tests required for exhaust-gas analysis, another test determines total hydrocarbons emitted from a non-operating vehicle. This is generally called the *shed test*. A vehicle is placed in a sealed enclosure called a "bag." The evaporating hydrocarbon emissions are then measured as they accumulate in the bag. The test is in two parts.

Simulated Sun Load—Emissions increase when the vehicle is heated because this creates fuel vapor. A parked vehicle being heated by sunlight is simulated by heating the gas tank at a linear rate. At the end of one hour, the hydrocarbon level in the enclosure is measured.

Hot-Vehicle Soak—This test simulates a parked vehicle that is still hot after being driven. The test vehicle is driven on the dynamometer until hot, then parked in the enclosure. At the end of a one-hour soak, the hydrocarbon level in the enclosure is measured again. Total hydrocarbon losses are determined by summing the data obtained from the two vehicle tests.

EMISSIONS-CONTROL EQUIPMENT

Since 1959, when the first emissions-control device was introduced, until the present, many emissions-control devices, hardware and systems have been devised by the automotive industry.

Competition among domestic manufacturers, pressures from foreign car companies and innovative ideas from outside suppliers or vendors have initiated improvements in engine designs to meet or exceed ever-tighter legislated emissions standards.

It would be impossible to describe all emissions-control equipment released on domestic vehicles during the the last three decades, or variations of each basic system. What follows describes and illustrates some of the more fundamental and common devices and systems.

POSITIVE CRANKCASE VENTILATION (PCV)

Crankcase emissions were the first target of lawmakers and automotive engineers because approximately one-third of total engine emissions originate from this point.

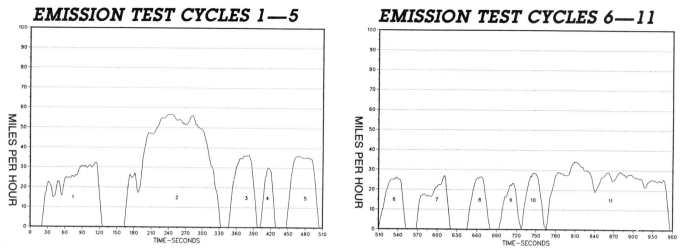

EMISSION TEST CYCLES 1—5

EMISSION TEST CYCLES 6—11

Current emissions testing requires pollutant measuring during 23 driving cycles. One difficult passing point is at 180 seconds. Vehicle is

By 1968, the PCV system was the standard crankcase ventilation method used on all domestic cars. It removes engine crankcase vapor resulting from normal engine *blowby*—unburned fuel and combustion products leaking past the rings into the crankcase. Manifold vacuum draws fresh air through the crankcase, which pulls the undesirable, corrosive gases and unburned fuel into the manifold so they can be burned in the engine. When the PCV valve is open, air flows through the air cleaner into the crankcase where it picks up vapor, then into the intake manifold.

Keep the PCV valve, filter and vent tube clean for correct engine operation because this is the only crankcase venting. Positive pressure in the crankcase can force oil out the main-bearing seals, oil sealing gaskets and can cause extra oil consumption.

PRE-HEATED AIR

This has long been a basic means of improving emissions, but was originally conceived to prevent carburetor icing.

Thermostatically controlled air cleaner intakes and a heated air source over the exhaust manifold are used to maintain intake air at approximately 100F (38C) or above. These supply warm air quickly during engine warmup and permit a leaner mixture, which reduces HC and CO.

Schematic of PCV system with valve open.

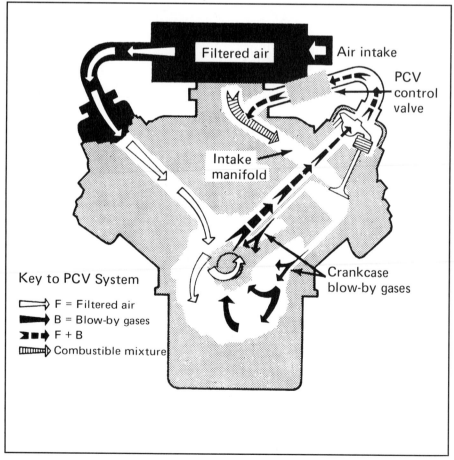

Key to PCV System

☐⟹ F = Filtered air
■➡ B = Blow-by gases
➤➤➡ F + B
▥➡ Combustible mixture

Filtered air

Air intake

PCV control valve

Intake manifold

Crankcase blow-by gases

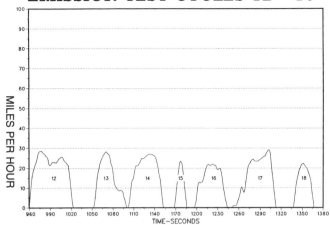

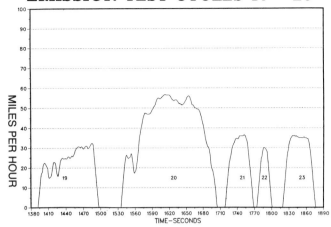

accelerated to over 50 mph just after choke is off and engine is not fully warm. Lots of opportunity for pollutants.

Typical 1970—'80s GM air cleaner package. Don't alter or modify, it works well.

Thermac control switch inside air cleaner is thermostatically actuated to open flapper valve and admit cooler air through snorkel.

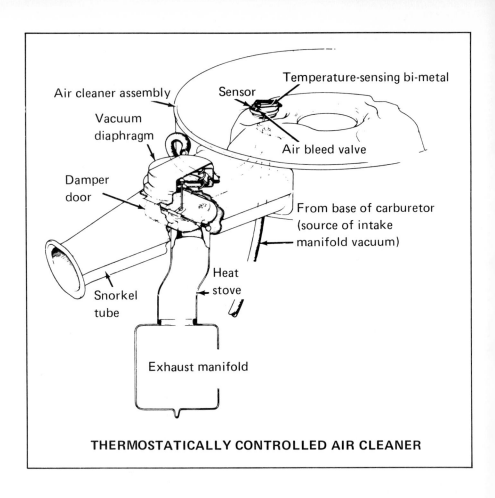

Temperature-sensing bi-metal

Sensor

Air cleaner assembly

Vacuum diaphragm

Damper door

Snorkel tube

Air bleed valve

From base of carburetor (source of intake manifold vacuum)

Heat stove

Exhaust manifold

THERMOSTATICALLY CONTROLLED AIR CLEANER

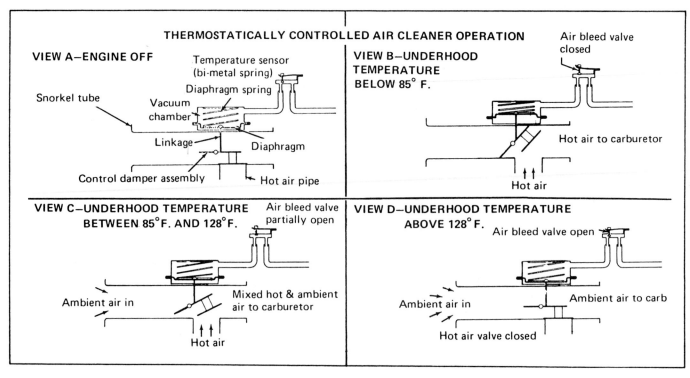

THERMOSTATICALLY CONTROLLED AIR CLEANER OPERATION

VIEW A—ENGINE OFF

Temperature sensor (bi-metal spring)

Diaphragm spring

Vacuum chamber

Snorkel tube

Linkage

Diaphragm

Control damper assembly

Hot air pipe

VIEW B—UNDERHOOD TEMPERATURE BELOW 85° F.

Air bleed valve closed

Hot air to carburetor

Hot air

VIEW C—UNDERHOOD TEMPERATURE BETWEEN 85°F. AND 128°F.

Air bleed valve partially open

Ambient air in

Mixed hot & ambient air to carburetor

Hot air

VIEW D—UNDERHOOD TEMPERATURE ABOVE 128°F.

Air bleed valve open

Ambient air in

Ambient air to carb

Hot air valve closed

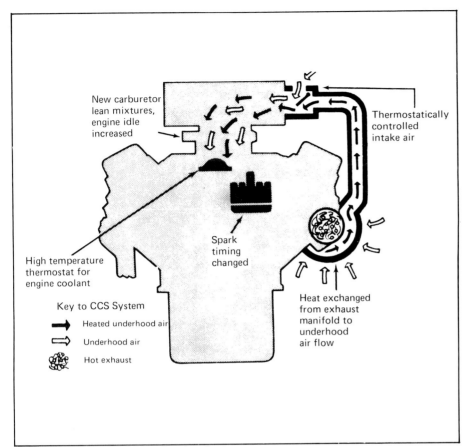

Schematic of Controlled Combustion System.

New carburetor lean mixtures, engine idle increased

Thermostatically controlled intake air

Spark timing changed

High temperature thermostat for engine coolant

Heat exchanged from exhaust manifold to underhood air flow

Key to CCS System
- Heated underhood air
- Underhood air
- Hot exhaust

Schematic of AIR system.

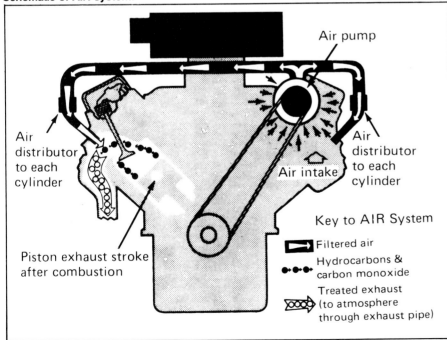

Air pump

Air distributor to each cylinder

Air distributor to each cylinder

Air intake

Piston exhaust stroke after combustion

Key to AIR System
- Filtered air
- Hydrocarbons & carbon monoxide
- Treated exhaust (to atmosphere through exhaust pipe)

The air cleaner assembly includes a temperature sensor, vacuum motor, control-damper assembly and connecting vacuum hoses. The temperature sensor controls the vacuum motor that operates the damper to regulate airflow to the carburetor. Preheated air from the exhaust-manifold heat tube, unheated underhood air, or a blend of the two, is passed to the carburetor.

EXHAUST CONTROLS

Various approaches have been taken to make engines meet exhaust-emissions standards. The first were engine and fuel-system modifications, air injection, and starting in 1971, EGR.

Engine Modifications—These were used from 1971—75 to meet emissions standards, with little regard to driveability and fuel economy. They included carburetor calibration and operational changes, and different distributor spark-advance and timing settings.

Carburetors were set with leaner mixtures at idle, off-idle and part-throttle. Mechanical limiters on idle-mixture screws prevented rich idle mixtures. Dashpots and throttle retarders and special idle-setting solenoids were used. Choke mixtures were eliminated shortly after the engine started. Inlet air was heated to ensure good fuel evaporation and distribution.

Distributors were set with more retard at idle and part-throttle, and various switches, valves and other controls were used to govern advance or retard as required to meet the standards.

Basic engine modifications were made. Compression ratios were reduced to lower combustion pressures and temperatures. Valve timing was altered and combustion chambers redesigned. Exhaust restrictors were also widely used.

It's difficult to sort out which modifications are doing what. But, let's oversimplify and say the retarded spark helps reduce HC emissions by increasing heat in the exhaust for complete burning of any remaining hydrocarbons. It also does not let the engine develop the peak pressures it is capable of, so NOx is reduced too.

Keeping the compression ratio down reduces flame travel speed and keeps the peak pressures from becoming high, so NOx is reduced. EGR accomplishes the same gross effect.

Air Injection Reactor (AIR)—The AIR system was used on most passenger-car engines by 1972. It consists of an air-injection pump, necessary brackets and drive belts, air-injection tubes (one for each cylinder), air-diverter valve, check valves, air-manifold assemblies, and the hoses necessary to connect the components.

Carburetors and *distributors* for cars equipped with an AIR are calibrated for those particular engines. Therefore, they should not be interchanged or replaced by a carburetor or distributor from engines without AIR systems.

The *air injection pump* compresses fresh air and injects it through the AIR manifold hoses and injector tubes into the exhaust system near the exhaust valves. The oxygen in the fresh air helps burn the unburned portion of the exhaust gases, thus minimizing exhaust contamination.

The *diverter valve,* when triggered by a sharp increase in manifold vacuum or deceleration, temporarily shuts off the injected fresh air to the exhaust ports, and prevents backfire during this richer period. At high engine speeds, excess air is dumped through a pressure-relief valve that is part of the diverter valve.

The *check valves* prevent exhaust gas from entering and damaging the air pump because back-flow can occur under many operating conditions.

When correctly installed and maintained, the AIR system will effectively reduce exhaust emissions. If the system doesn't function, or is removed, CO emissions are high because the carburetor supplies a rich mixture to combine with the injected air.

Because the state of engine tune influences the amount of unburned gases in the exhaust, it has a direct bearing on AIR system operation. Be sure the engine is tuned correctly before analyzing system problems.

The basic test to determine if the AIR is working is to loosen the hose clamp at the check valve. Pull the hose from the valve. Start the engine and hold your hand over the end of the hose. If the hose blows air with noticeable force, it is operating. Revving the engine should increase the pressure. Don't run the engine long with the hose disconnected because exhaust against the backside of the check valve may eventually damage it. If backfiring and popping noises occur when slowing to a stop, the diverter valve needs attention.

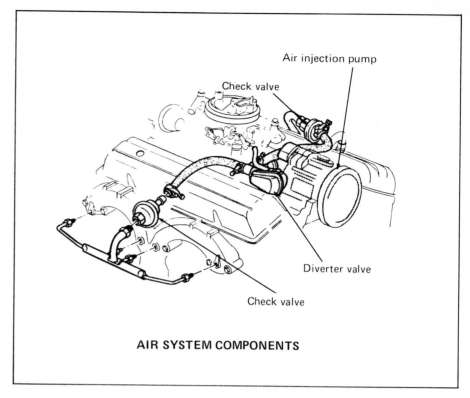

AIR SYSTEM COMPONENTS

AIR system operation.

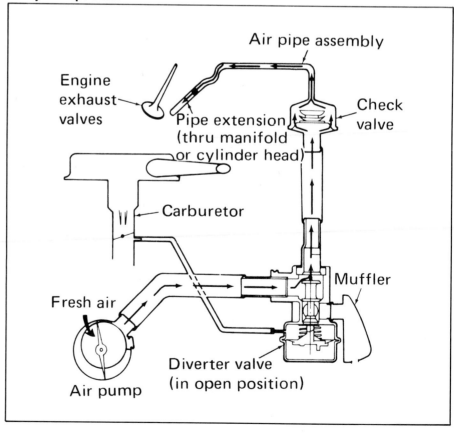

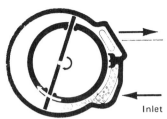 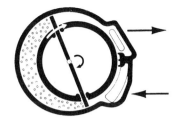 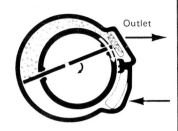

The vane is travelling from a small area into a larger area—consequently a vacuum is formed that draws fresh air into the pump.

As the vane continues to rotate, the other vane has rotated past the inlet opening. Now the air that has just been drawn in is entrapped between the vanes. This entrapped air is then transferred into a smaller area and thus compressed.

As the vane continues to rotate it passes the outlet cavity in the pump housing bore and exhausts the compressed air into the remainder of the system.

AIR INJECTION PUMP OPERATION

Diverter valve operation.

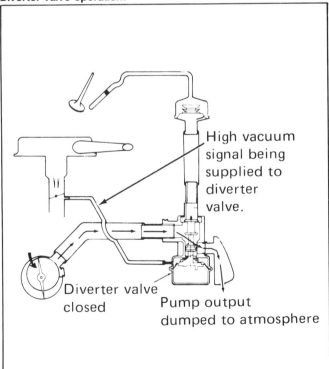

High vacuum signal being supplied to diverter valve.

Diverter valve closed

Pump output dumped to atmosphere

Diverter valve components.

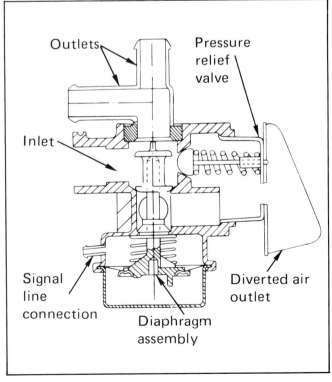

Outlets

Pressure relief valve

Inlet

Signal line connection

Diaphragm assembly

Diverted air outlet

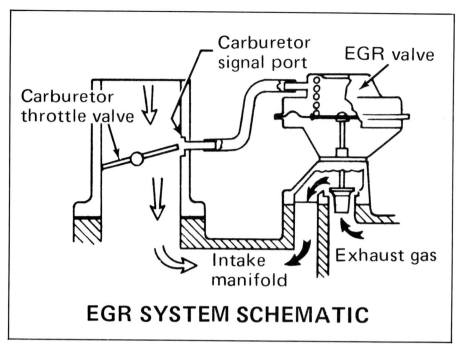

EGR SYSTEM SCHEMATIC

Carburetor throttle valve

Carburetor signal port

EGR valve

Intake manifold

Exhaust gas

EGR reduces NOx by lowering peak pressures and temperatures in combustion chambers. EGR accounts for as much as 15% of intake mixture volume.

Exhaust Gas Recirculation (EGR)—
Nitrogen tends to combine with oxygen at 2040F (1116C) and form NOx. Engine combustion temperatures often exceed this value. To reduce the formation of NOx, an EGR valve meters inert exhaust gases into the intake manifold. These displace some of the A/F mixture in the cylinders so peak-combustion temperatures and pressures are lowered. A portion of the exhaust gas is being recirculated into the combustion chamber, hence the term *exhaust gas recirculation*.

EGR doesn't affect vehicle driveability at idle or WOT when no EGR is occurring. But engine surging and uneven running happen when recirculation amounts to 15% of mixture volume. Poor driveability is most apparent at light-throttle/high vacuum running at speeds from 25—60 mph.

The *EGR valve* mounts on the intake manifold and connects to an exhaust-manifold port. The valve is closed during idle and deceleration to prevent rough running caused by excessive exhaust-gas dilution of the mixture. It gets its opening signal from the carburetor. Vacuum from an off-idle port opens the valve when the throttle blade is opened past idle.

A second (upper) EGR port bleeds air into the vacuum channel to reduce the signal supplied by the lower port. This times the EGR valve to the port's location in the carburetor bore and the throttle opening.

As the throttle is opened farther into the part-throttle range, the upper port ceases to function as a bleed. It is exposed to manifold vacuum, supplements the vacuum signal from the lower port, and the EGR valve is maintained in the desired position. High vacuum opens the valve wide, low vacuum opens it slightly. Below 5-in.Hg vacuum or at WOT, the valve closes.

It is closed at idle to avoid excessively diluting the mixture because dilution already exists at idle from residual exhaust products in the cylinders. It is closed at WOT to avoid reducing the available power from the engine at heavy loads. The single-diaphragm EGR responds to throttle position and airflow through the engine.

Dual-diaphragm EGR valves were introduced in 1974 to improve exhaust gas metering into the intake manifold according to engine speed and load. This valve monitors and responds to intake-manifold vacuum as an indicator of engine load. The result is improved driveability with less

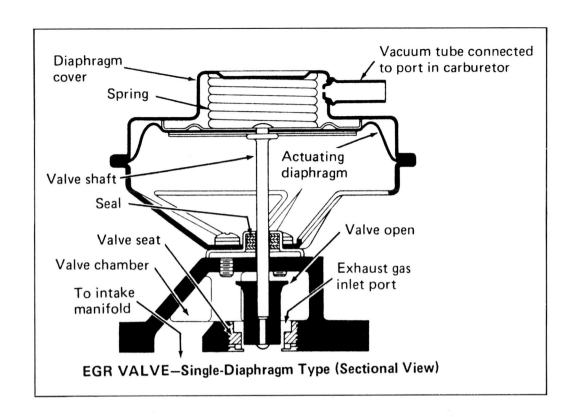

EGR VALVE—Single-Diaphragm Type (Sectional View)

Engine surging at light load and part throttle is minimized by dual-diaphragm EGR valve.

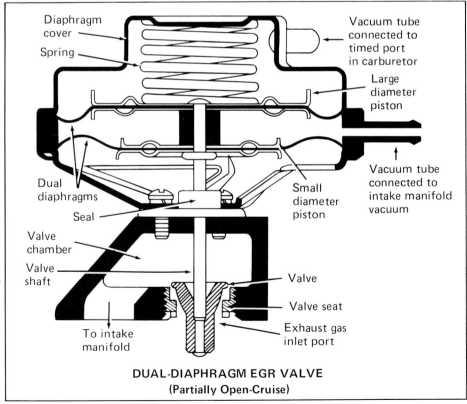

DUAL-DIAPHRAGM EGR VALVE
(Partially Open-Cruise)

surging at part-throttle.

An advantage of the dual-diaphragm control can be seen by reviewing a worst-case. A vehicle is traveling at medium speed with part-throttle on a gentle downward grade. With a single-diaphragm unit, the throttle valve would be positioned near the timed port, and the EGR valve would be completely open. Too much exhaust dilution of mixture, surging, or uneven running would result.

With dual-diaphragm control, under the same road condition, timed-port vacuum tends to open the EGR valve, but the high manifold vacuum tends to close it. The result will be a partially open valve metering a tolerable amount of exhaust gas into the engine.

Many people pull the vacuum hose off the EGR to stop it from working. Be aware that extra vacuum advance is set in cars with EGR valves. When the EGR is not functioning, the mixture to the cylinder is not diluted and damaging detonation is likely.

EVAPORATIVE-EMISSIONS CONTROLS

Evaporative emissions include hydrocarbons in fuel spilled or evaporated from fuel tanks and carburetors. You might consider them to be an insignificant part of overall emissions, but these are estimated to equal the amount that would occur if crankcase ventilation were uncontrolled.

The gas tank and carburetor were usually vented to the atmosphere. Venting got rid of vapor and helped hot-starting capabilities. There was no real concern about emissions caused by spilled fuel during overfilling or gasoline sloshed out of vents during sudden maneuvers. In 1970—71, total evaporative emissions allowed from any car were 6 grams per hot-soak. In 1972, the allowance was reduced to 2 grams per hot-soak.

EVAPORATIVE CONTROL SYSTEM (ECS)

This system reduces fuel-vapor emissions that would otherwise vent to the atmosphere from the gasoline tank and carburetor bowl. The ECS is one emission-control device that actually aids vehicle

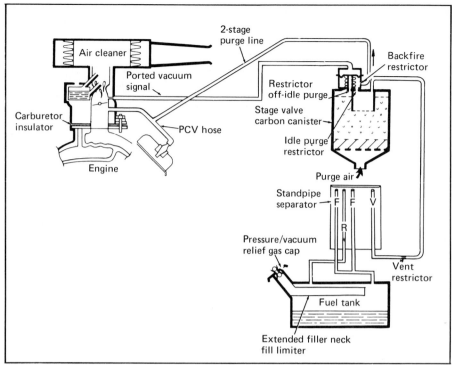

Evaporative Control System (ECS).

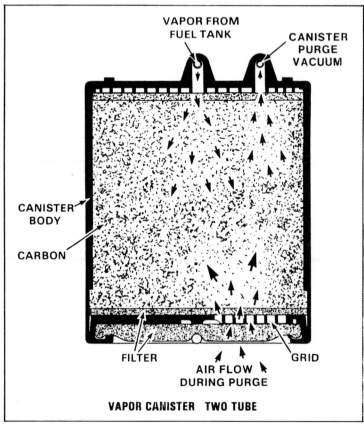

Drawing courtesy Rochester Products Division, GM.

168

efficiency. Anytime raw fuel can be burned instead of polluting the air, economy and ecology are served.

The obvious part of this system is the charcoal canister usually found in the engine compartment. The canister contains charcoal granules. The fuel-vapor inlets and outlets are on one side of the charcoal and the purge-air opening and filter are on the other. *Purge air* is air drawn through the saturated charcoal to extract stored fuel vapor.

Vapor enters the canister from the carburetor bowl vent and the fuel-tank liquid/vapor separator. It is absorbed onto the surfaces of the charcoal granules and stored in the air space in the canister. When the car is restarted, fuel vapor and fresh air are drawn into the carburetor or intake manifold to be burned in the engine. This is controlled by a timed port, so purging the canister only occurs when the engine can consume the vapor.

Charcoal Canister Design—The canister contains 300—625 grams of activated charcoal (carbon) in granules with 0.033—0.094-in. diameter. Each gram of charcoal can hold up to 35% of its weight in fuel vapor. Fuel vapor molecules are condensed and held onto the surface of the charcoal. Vapor molecules are easily dislodged by flowing fresh air through the charcoal bed during purge cycles.

Canister purging methods vary. Usually there are two calibrated orifices, either in the purge valve canister or carburetor throttle body, to return vapors to the engine. The smaller of the two connects directly to the intake manifold and passes fuel vapors or purge air any time the engine runs. This is called *constant bleed purge.* The small orifice is sized according to how much vapor the engine can tolerate at idle.

A second orifice, also connected to the manifold or carburetor, is controlled by a vacuum-operated valve on the canister and a timed port (off-idle) in the carburetor. This increases purging at speeds above idle when engine speed is sufficient to accept the additional fuel/vapor.

GM used a thermal-delay valve on some models so the canister was not purged until the engine reached operating temperature. Purging at idle or with a cold engine causes rough running and increased emissions because the vapor added to the manifold can cause an over-rich mixture.

As standards tightened, changes in evap-orative emissions control were required. The two-valve canister was replaced during the '80s by a simpler two-tube canister. It was widely used by GM during 1981—86.

Fuel-Supply Design—Fuel-tank filling and construction have been improved for the ECS. Thermal expansion is accounted for by trapping up to 3 gallons of air during tank filling. It is allowed to escape to the top of the main tank, where it vents to the canister through a liquid/vapor separator. Any liquid trapped in the separator(s) drains back to the tank and the vapor passes to the canister.

The gas cap is designed to contain vapor in the tank and force it through the separator and into the canister. It takes approximately 1/2 psi to make the canister system work. Under extreme conditions, when the tank pressure goes above the cap's pressure limit, it vents to atmosphere.

First-design caps would blow off at 1—2 psi pressure. Starting in 1972, the fuel-tank caps had a higher pressure and the tank, straps and related components were made stronger to withstand this increase. The caps also included a vacuum vent to release any vacuum forming in the tank as fuel is used.

Do not mix gas caps among vehicles. With its large surface area, it doesn't take much pressure to damage or burst a fuel tank. *Any* emissions control components that fail should be replaced with OEM parts to guarantee correct performance.

EMISSIONS-CONTROL EVOLUTION

EARLY DESIGN COMPROMISES

During the first years of meeting HC, CO and NOx standards, the automobile industry made undesirable compromises in engine design and calibration. These had detrimental effects on engine performance, driveability and fuel economy.

The low point was reached during 1974. Drivers suffered with vehicles that stalled after cold-start due to lean chokes, surged on the road due to lean part-throttle calibration and reduced spark, and had poor fuel economy due again to retarded spark.

EMISSIONS-CONTROL IMPROVEMENTS

Improved technology released engine designers from such engineering com-promises. First in 1975, and again in 1981, the automobile industry made quantum jumps in improving driveability and fuel economy while still meeting more stringent emissions standards.

Catalytic Converter—In 1975, the industry introduced the *catalytic converter,* which transferred the burden of reducing contaminants from within the engine to a filter in the exhaust stream.

Engineers were able to return to reasonable automatic-choke calibrations for satisfactory cold-starts and warmup. Carburetor A/F ratios were richened for surge-free part-throttle operation, and spark advance settings were tailored for good fuel economy and engine performance.

The converters used in most vehicles from 1975 have two configurations: *beaded* and *monolithic.* Several catalysts can oxidize hydrocarbons in the exhaust system, changing them to harmless water vapor or carbon dioxide. The most widely used catalyst is built with platinum, which is deposited on the surface of thousands of granular beads (beaded), or the surface of monolith honeycomb designs (monolithic). The catalyst container looks like a muffler, and is placed as close to the engine as possible to minimize heat loss.

The first catalysts were the *single-bed design,* used only to reduce HC and CO, with NOx reduction handled exclusively by EGR. During the early '80s, as limits for NOx were drastically lowered, *dual-bed* convertors were released. One bed for HC and CO, enhanced by introducing controlled air injection into the bed; a second bed used a rhodium catalyst to oxidize the NOx.

Emissions Label—All manufacturers are required to install a "Vehicle Emission Control Information" label in the engine compartment. The label specifies:
- Engine size and type.
- Engine timing.
- Spark plug gap.
- Curb- and fast-idle settings.
- Setting instructions.
- An emission hose routing schematic.

No matter how effective or elaborate the emissions control system, a weak spark, fouled plug, cracked distributor or many other engine items out-of-tune can hamper system efficiency. Fuel economy, emissions control, and driveability all suffer if an engine is not kept in tune.

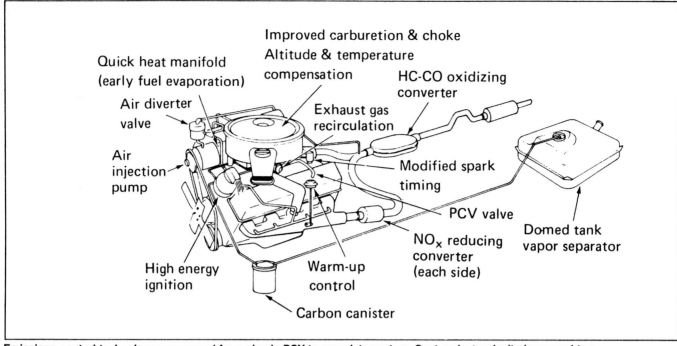

Quick heat manifold
(early fuel evaporation)

Air diverter
valve

Air
injection
pump

High energy
ignition

Improved carburetion & choke
Altitude & temperature
compensation

Exhaust gas
recirculation

HC-CO oxidizing
converter

Modified spark
timing

PCV valve

Warm-up
control

NO_x reducing
converter
(each side)

Carbon canister

Domed tank
vapor separator

Emissions control technology progressed from simple PCV to complete system. Cost and complexity increased too.

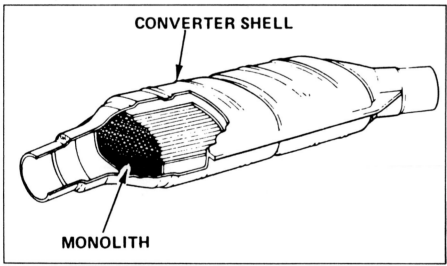

CONVERTER SHELL

MONOLITH

Single-bed monolithic catalytic converter. Drawing courtesy GM.

Microprocessor Control—Applying dual-bed catalytic convertors to engines was made possible by the second big leap forward in 1981 by the automobile industry. They incorporated microprocessors (computers) into controlling fuel metering and emissions.

An automobile microprocessor is supplied information including, but not limited to: engine rpm, car speed, engine load (based on manifold vacuum and carburetor throttle positions), engine coolant temperature, exhaust oxygen-content and barometric pressure.

It then processes this information and manipulates engine functions previously done mechanically. These include, but again are not limited to: carburetor A/F ratio, spark advance, idle speed, EGR flow and timing, injection sequence timing and direction of flow, transmission lock-up speeds and canister purge rate.

This book doesn't describe the theory or inner workings of a microprocessor. But I can state with confidence that such devices have enabled design engineers to control engine settings and specifications more accurately and quickly than heretofore considered possible. Microprocessors in automobiles have proven to be durable, consistent, and relatively trouble-free for many

Dual-bed monolithic catalytic converter. Drawing courtesy GM.

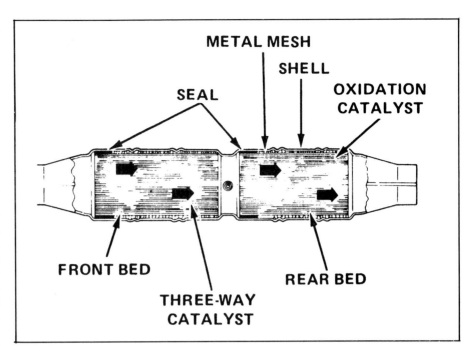

miles. Most car owners are unaware their car has them.

Using microprocessors caused the design of new emissions devices. These include: electronic-air-management divert and switching valves, electric solenoids that control EGR flow sequence and timing, fuel metering control systems, and canister purge flow and timing. New electronic distributors permit the microprocessor to control spark advance and accurately meet all engine timing requirements, as well.

The microprocessor has enabled the carburetor's mixture to be closely maintained at the ideal A/F ratio (stoichiometric, 14.7:1). This is the A/F ratio window in which CO and HC are 85—90% converted or oxidized, and NOx is 85—90% reduced in their respective catalytic-converter beds.

Accurate timing and switching of the air pump is now possible with microprocessor-controlled solenoids. Control is built into one unit, aptly named the Electric-Divert/Electric-Switching Valve. Two types are available: one using manifold vacuum to control the diverting function, the other using air pressure from the pump. Construction and hose routings are simplified.

Basically, the system supplies additional air to the exhaust ports to reduce HC and CO. The microprocessor controls all operation, based on program instructions written by design engineers. During the cold or "open loop" phase, air flows to the exhaust ports. When the system is switched into the "closed loop" phase, the diaphragm chamber is vented to atmosphere, and airflow is directed between the convertor beds to make an oxidizing atmosphere for better emissions control. Under certain conditions, such as WOT, the supplementary air is diverted overboard (externally) either into the air cleaner or a silencer.

Thus, driveability, performance, fuel economy and emissions control have been improved by applying microprocessors to carburetion. Functions that were controlled mechanically are now governed digitally.

Single-bed beaded catalytic converter. Drawing courtesy GM.

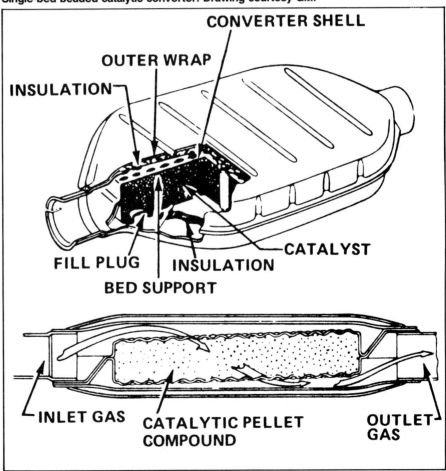

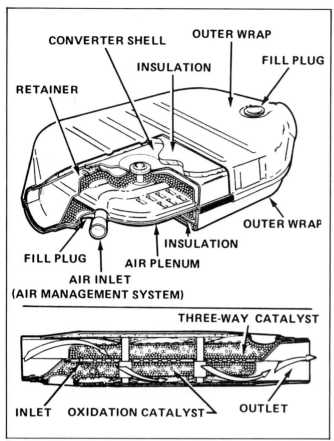

CONVERTER SHELL

OUTER WRAP

FILL PLUG

INSULATION

RETAINER

OUTER WRAP

INSULATION

FILL PLUG

AIR PLENUM

AIR INLET
(AIR MANAGEMENT SYSTEM)

THREE-WAY CATALYST

INLET OXIDATION CATALYST OUTLET

Dual-bed beaded catalytic converter. Drawing courtesy GM.

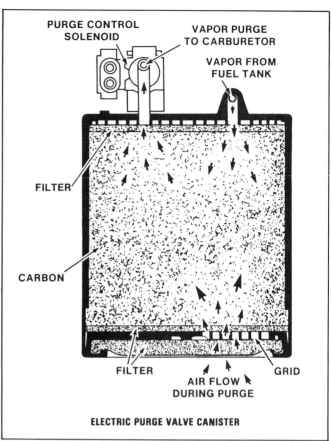

PURGE CONTROL
SOLENOID

VAPOR PURGE
TO CARBURETOR

VAPOR FROM
FUEL TANK

FILTER

CARBON

FILTER

GRID

AIR FLOW
DURING PURGE

ELECTRIC PURGE VALVE CANISTER

ECM energizes purge control solenoid to close off purge air or de-energizes it to open airflow passage through canister. Drawing courtesy Rochester Products Division, GM.

Electric Divert/Switching valve contolled by ECM either allows air pump to flow air to exhaust ports (open loop, cold engine), or sends air between catalytic converter beds (closed loop, warm engine). At WOT, air is diverted externally (into air cleaner). Drawing courtesy Rochester Products Division, GM.

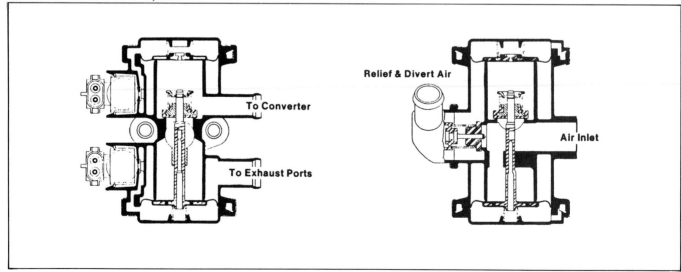

To Converter

To Exhaust Ports

Relief & Divert Air

Air Inlet

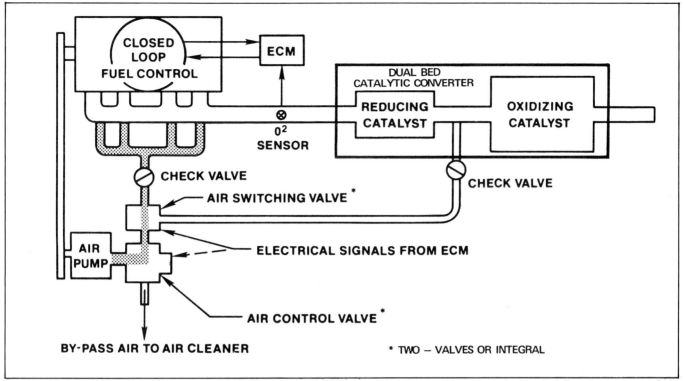

Fresh air from air pump is directed by signals from ECM to exhaust ports during open loop, cold engine operation. Drawing courtesy GM.

Fresh air from air pump is sent between catalytic converter beds during closed loop, warm engine operation. This supplies additional oxygen to catalysts and decreases CO, HC and NOx levels. Drawing courtesy GM.

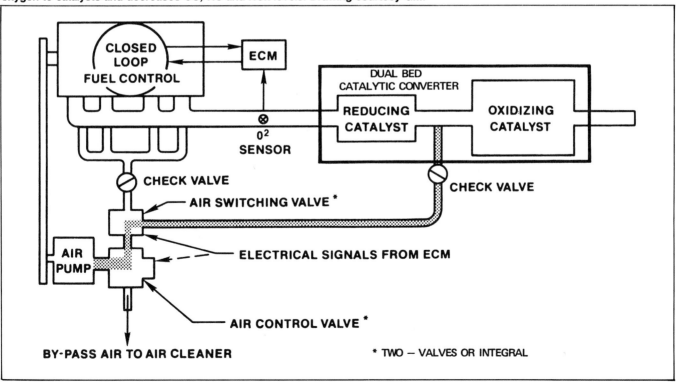

METRIC CUSTOMARY-UNIT EQUIVALENTS

Multiply:		by:		to get:	Multiply:		by:		to get:

LINEAR

inches	X	25.4	=	millimeters(mm)	X	0.03937	=	inches
miles	X	1.6093	=	kilometers (km)	X	0.6214	=	miles
inches	X	2.54	=	centimeters (cm)	X	0.3937	=	inches

AREA

| $inches^2$ | X | 645.16 | = | $millimeters^2 (mm^2)$ | X | 0.00155 | = | $inches^2$ |
| $inches^2$ | X | 6.452 | = | $centimeters^2 (cm^2)$ | X | 0.155 | = | $inches^2$ |

VOLUME

| quarts | X | 0.94635 | = | liters (l) | X | 1.0567 | = | quarts |
| fluid oz | X | 29.57 | = | milliliters (ml) | X | 0.03381 | = | fluid oz |

MASS

pounds (av)	X	0.4536	=	kilograms (kg)	X	2.2046	=	pounds (av)
tons (2000 lb)	X	907.18	=	kilograms (kg)	X	0.001102	=	tons (2000 lb)
tons (2000 lb)	X	0.90718	=	metric tons (t)	X	1.1023	=	tons (2000 lb)

FORCE

| pounds−f(av) | X | 4.448 | = | newtons (N) | X | 0.2248 | = | pounds−f(av) |
| kilograms−f | X | 9.807 | = | newtons (N) | X | 0.10197 | = | kilograms−f |

TEMPERATURE

Degrees Celsius (C) = 0.556 (F - 32)   Degree Fahrenheit (F) = (1.8C) + 32

°F   -40   32   98.6   212   °F
     0   40 80   120   160 200   240   280   320

°C   -40   -20   0   20   40   60   80   100   120   140   160   °C

ENERGY OR WORK

| foot-pounds | X | 1.3558 | = | joules (J) | X | 0.7376 | = | foot-pounds |

FUEL ECONOMY & FUEL CONSUMPTION

| miles/gal | X | 0.42514 | = | kilometers/liter(km/l) | X | 2.3522 | = | miles/gal |

Note:
235.2/(mi/gal) = liters/100km
235.2/(liters/100km) = mi/gal

PRESSURE OR STRESS

inches Hg (60F)	X	3.377	=	kilopascals (kPa)	X	0.2961	=	inches Hg
pounds/sq in.	X	6.895	=	kilopascals (kPa)	X	0.145	=	pounds/sq in
pounds/sq ft	X	47.88	=	pascals (Pa)	X	0.02088	=	pounds/sq ft

POWER

| horsepower | X | 0.746 | = | kilowatts (kW) | X | 1.34 | = | horsepower |

TORQUE

pound-inches	X	0.11298	=	newton-meters (N-m)	X	8.851	=	pound-inches
pound-feet	X	1.3558	=	newton-meters (N-m)	X	0.7376	=	pound-feet
pound-inches	X	0.0115	=	kilogram-meters (Kg-M)	X	87	=	pound-inches
pound-feet	X	0.138	=	kilogram-meters (Kg-M)	X	7.25	=	pound-feet

VELOCITY

| miles/hour | X | 1.6093 | = | kilometers/hour(km/h) | X | 0.6214 | = | miles/hour |

INDEX